Histological Techniques
An Introduction for Beginners in Toxicology

Histological Techniques
An Introduction for Beginners in Toxicology

Robert Maynard
University of Birmingham, UK
Email: robertmaynard3@gmail.com

Noel Downes
Sequani Limited, UK
Email: noel.downes@sequani.com

Brenda Finney
Sequani Limited, UK
Email: brenda.finney@sequani.com

Paperback ISBN: 978-1-83916-147-6
Hardback ISBN: 978-1-84973-992-4
EPUB ISBN: 978-1-78262-533-9

A catalogue record for this book is available from the British Library

The Royal Society of Chemistry is a charity, registered in England and Wales, Number 207890, and a company incorporated in England by Royal Charter (Registered No. RC000524), registered office: Burlington House, Piccadilly, London W1J 0BA, UK, Telephone: +44 (0) 20 7437 8656.

Visit our website at www.rsc.org/books

Preface

This little book is for toxicologists who are in the early stages of their careers and who want to know something about histological techniques. I have assumed that the reader has no experience of histological methods but wants to use them in his or her research. I have thus followed at least the first part of the advice given to the White Rabbit by The King of Hearts: "begin at the beginning, and go on till you come to the end, then stop". I have not gone on to the end! Other authors have done so, or tried to do so, and their works are listed in the bibliography.

No book on histological methods can claim to be entirely original and this one is certainly not. What is perhaps unusual is the advice on how to go about histological research and how to look at histological sections. Such advice is seldom provided in books on techniques. Whether my advice will be useful to the reader remains to be seen. I have also tried to include "tricks of the trade", which I picked up from better histologists and histology technicians. Such tricks and dodges can make all the difference to the beginner: they can prevent disappointment.

I began writing this book alone but it rapidly became obvious that I needed help. This was offered by Noel Downes and his colleague Brenda Finney, who have joined me as co-authors.In addition to much improving the text Noel Downes has contributed all the photomicrographs and the majority of the chapter dealing with histochemistry. Brenda Finney has contributed on immune histochemistry and has

Histological Techniques: An Introduction for Beginners in Toxicology
By Robert Maynard, Noel Downes and Brenda Finney
© Maynard, Downes and Finney 2020
Published by the Royal Society of Chemistry, www.rsc.org

improved the text in other chapters. Without their involvement, this book would not have been completed.

The debt I owe to other writers on histological techniques will be obvious to the reader. Dr John Findlay read and commented on the chapter on microscopy: I have benefitted from his expert advice. Dr Peter Evennett also read a part of this chapter and provided comments that proved to be critical and helpful. I should also like to acknowledge the advice I received early in my career from Professor Fritz Jacoby (then recently retired from his post of Professor of Histology at University College Cardiff) and from Mr Les Jones (then Chief Technician, Department of Anatomy, University College Cardiff), who was an outstanding exponent of classical histological techniques. Drs Marshall Craigmyle and John Findlay taught me histology: I remain grateful to them.I should also like to acknowledge those two distinguished histologists from Oxford University: the late Dr J R Baker FRS and the late Dr H M Carleton.Their works have been a constant source of inspiration.

I must also thank those authors and publishers who have given permission for material to be quoted in this book. Every effort to trace original sources has been made.

I must also thank the staff at the Royal Society of Chemistry for seeing this book through the press with dispatch.

Finally, I should like to acknowledge my wife, Sandra, who, as an artist, has listened to me talking about microscopes, sections, stains and colours for many years. Her patience has been unfailing.

R L Maynard
Devon

Contents

Histological Techniques: An Introduction for Beginners in Toxicology
By Robert Maynard, Noel Downes and Brenda Finney
© Maynard, Downes and Finney 2020
Published by the Royal Society of Chemistry, www.rsc.org

CHAPTER 1

Introduction

Histology is the science of cells and tissues and plays an essential part in toxicological investigations. Though histological work is often limited to specialists in the subject there is no reason why this should be so: the application of basic histological methods is within the reach of any research worker in the biological sciences. Routine histopathology can be a quick and cheap way to answer simple questions. Many questions are initially relatively straight-forward and histopathology is very good at guiding the researcher towards the specifics where more in depth investigations can be better targeted.

Planning research is all about asking questions and devising means for answering them. In this chapter the reader is introduced to the sort of problems which are amenable to histological approaches. Histology is often regarded as a descriptive discipline and so, to a large extent, it is. But histology can be used to test hypotheses and thus can make a great contribution to the advance of biological research. Linking histological techniques with how to think about histological problems is the aim of this introductory chapter.

1.1 HISTOLOGY FOR TOXICOLOGISTS

This chapter sets out the background to histological studies in toxicology and takes the reader into a few imaginary studies. These

Histological Techniques: An Introduction for Beginners in Toxicology
By Robert Maynard, Noel Downes and Brenda Finney
© Maynard, Downes and Finney 2020
Published by the Royal Society of Chemistry, www.rsc.org

provide a fore-taste of subsequent chapters and allow the beginner to see what an interesting and scientific subject histology actually is.

The study of toxicology includes the study of damage to cells and tissues produced by chemicals. Such damage may be studied in a variety of ways; for example, damage to the lung or liver may be studied by tests of the functional capacities of these organs. Lung function tests and liver function tests provide physiological and bio-chemical information about how well these organs are working. Damage may also be studied by histological methods: tissues are examined for signs of damage. In some cases evidence of structural damage can be related to evidence of impairment of function. Evidence of structural damage also provides clues as to the mechanisms of the effect of chemicals. Some chemicals produce very specific forms of tissue damage and these have been described in detail. Textbooks of toxicology contain detailed descriptions of, for example, the effects of compounds such as chloroform or carbon tetrachloride on the liver.

A search for evidence of structural damage is an essential part of the study of the potential toxicological effects of any new chemical. Standard approaches have been developed and production-line methods are applied to sampling of tissues, processing of tissues for histological examination and to the examination itself of the histo-logical sections produced. These production-line methods allow the screening of a wide range of tissues and are carried out to Good Laboratory Practice standards. Such standards ensure that the results of the studies are reliable.

Histological techniques are also applied, but in a less standardised way, during research into the effects of chemicals. For example, one might be studying the efficacy of a skin decontaminant and become concerned about the possible toxicological effects of the compound on the skin. It would be entirely reasonable, indeed necessary, to examine samples of skin histologically in an attempt to find evidence of damage. Similarly, if one were studying the effects of nano-particles on the lung one would wish to look closely at the lungs of animals killed at intervals after exposure to the material under study. Such work is a little different from the production-line approach taken in the screening of chemicals for possible toxicological effects. It tends to be done on a smaller scale and, perhaps, sometimes by less experienced workers. Of course such work should be performed to high standards and in accordance with the guidance of Good Laboratory Practice and Health and Safety standards. But varying and extending the approach as new ideas occur is the essence of research and producing a definitive plan of the methods to be used before any methods have been deployed is often impossible. For example, one might begin by examining paraffin sections of kidney and then decide that frozen sections that allowed the histochemical study of specific enzymes were needed. This "following the clues as they appear" approach is essential in research work. However, there are dangers in such an approach. One might be led to inappropriate methods or one might be led to difficult methods of which one has little experience and which require great skill before the results can be accepted as reliable. The "try 'em all" approach is ever-appealing and needs to be resisted.

Workers with limited experience of histological methods are often tempted to "have a look" at the tissue just in case anything odd or interesting can be seen. In an ideal world such work would be undertaken by an experienced histopathologist. This, however, is far from an ideal world and histopathologists are neither many nor under-employed. Thus, the toxicologist might feel he or she should "do some histology" as a part of their research project. Such an aspiration should be encouraged: much can be learnt from histological studies. But, of course, badly performed histological studies, like all badly performed studies, may be worse than useless in that they can be misleading. Poor tissue preparation may lead to incorrect appreciations of effects, or effects might be missed. Additionally, the wide range of artefactual changes that appear in badly prepared material may mislead the observer. The results of such studies may be disappointing and the researcher might be discouraged.

Toxicologists usually receive some training in histology and histopathology. Post-graduate courses in toxicology (often leading to Masters' degrees) usually include teaching on histopathology. But such courses cannot provide much practical experience and lessons learnt can be forgotten. Toxicologists with a predominantly biochemical or chemical background may lack familiarity with histology: this can hardly be regarded as their fault and they may have passed their training in histopathology with a sigh of relief and a resolution not to return to it. Toxicologists with a medical or veterinary background might be expected to be more enthusiastic about the study of histology and its derivative histopathology, though their experience of the practical aspects of the subjects may also be limited. The teaching of histology in medical schools has been squeezed over the last thirty years by the accretion of newer subjects to the curriculum. The traditional histologist might well complain that making drawings of sections has disappeared; he might despair that the examination of histological sections is fast following the practice of drawing! No medical students are now required to stain histological sections, let alone to cut sections. Are medical students required to examine histological sections with a microscope? In some medical schools this is not required and video presentations have taken the place of microscopy. Is this a bad thing? Most histologists would answer, yes. Nothing beats examining sections. Video presentations are better than relying only on an atlas with perhaps not more than a few photomicrographs of any one tissue, but they cannot replace the microscope in terms of showing the student the range of appearances of sectioned tissue. In addition, the resolution provided by a television screen may be poorer than that provided by a properly adjusted microscope. Veterinary students are better trained than medical students in the practical techniques of pathology, but in both professions it is accepted that histopathology is, in general, a postgraduate discipline.

1.2   SO WHO IS THIS BOOK FOR?

This book is aimed at the beginner, the scientist, and perhaps the post-graduate student, who makes no pretence of yet being a histologist, but who would like to know something more about the subject and would like to use histological techniques in his or her research but is not sure which techniques might be useful. It is not aimed at the medical or veterinary histopathologist; indeed, such experts might well treat the advice provided in this book with appropriate disdain.

Emphasis has been placed on both the "WHY" and the "HOW" of histological techniques. Many books describing histological methods are available. Such books offer the beginner a vast, indeed bewildering, range of techniques and this may be off-putting or even confusing. Books explaining why specific methods might be used are fewer: textbooks of histochemistry, for example, explain in detail how specific enzymes might be studied but, surprisingly, provide little advice on why one might wish to undertake such studies in the first place. Diagnostic histochemistry is a little better served: those interested in diseases of the liver and, perhaps specifically, in the formation of bile, or in tumours that produce certain hormones, might expect useful histochemical methods to be described within accounts of these diseases.

What are the problems likely to be faced by our ideal reader? Much depends of the environment in which he or she works. If a large, well-staffed, helpful and not over-loaded histopathology service laboratory, staffed with expert technicians, is available to process our readers' samples, and if expert histopathologists are accessible and willing to teach and discuss results, then our reader might feel that he or she is wasting time in trying to learn about histology by the "do it yourself" method. Nothing could be further from the truth. The better one's understanding of the methods used, the better one's under-standing of the results produced and the reports provided by the experts. On the other hand, there might be little support available from experts and the scientist might need to prepare his own material and examine it him- or herself. Whilst this might be deprecated by experts, it is in fact by far the best approach. One need not strive to become an expert histopathologist with a wide knowledge of tissue reactions and an ability to detect and classify innumerable variants of rare tumours: what we are interested in is what happens in "our" tissue when it is exposed to "our" chemical. Thus, there is nothing to prevent someone who is not a trained histopathologist becoming a real expert on the effects of skin-decontaminant X on the sweat glands or melanocytes of the skin. Specialised expertise, acquired by practice, is difficult to beat. The personal satisfaction of knowing more about one's own area than the universal expert is also far from negligible.

One of the greatest of histologists, Manfred Gabe, wrote in 1968:

"The young post-graduate embarking on a career of research has often been lulled for years on global attitudes. In nine cases out of ten he imagines that the primordial condition for success in scientific work is aptitude for philosophical generalization. Ready to juggle with fundamental ideas and edify grand hypotheses, our post-graduate would at the most deign to perform an experimentum crucis but would reject indignantly as a basely material task the very idea of having to handle a microtome, pipette or precision balance. Those of our future scientists who turn towards histology dream of sitting be-fore an excellent microscope and examining preparations that have been made, cleaned and labelled by technicians to finally produce a theory of the structure of living matter."

This is, perhaps, an exaggeration or perhaps not! Gabe's book *Histological Techniques* ran to 1100 pages: the results of a survey to discover how many research workers using histological techniques are acquainted with the work would be interesting. If we judge by comparing Gabe's trenchant criticism of "haematoxylin and eosin" (H&E) as a histological method with the universal adoption of this method, to the exclusion of nearly all others, then the number is likely to be low. The number, including the present author, who have not perused Gabe's atlas of the histology of that elusive New Zealand reptile, the tuatara, must be very large.

1.3   WHAT IS HISTOLOGY AND WHAT IS HISTOPATHOLOGY?

Histology is the scientific discipline that involves the study of the structure of normal cells, tissues and organs. The essential instrument is the microscope. Histopathology involves the study of the changes in cells, tissues and organs that occur as a result of disease processes or insults, for example, the effects of physical or chemical injury. Again, the microscope is the essential instrument. Of course, when "the microscope" is mentioned we imply all the variants of techniques that are included in the term "microscopy", and thus we should consider light microscopy in all its many forms and electron microscopy, again in its many forms. The changes sought by histopathologists are changes that can be seen, but making the changes visible often requires the application of a range of techniques. At its simplest level, histological sections are prepared and examined. Their appearance is compared with that of normal tissue, either by reference to controls or to one's memory of the normal appearance of the tissue being studied. Identifying visible changes is the first link in a chain of intellectual effort that includes deductions about the effects of the changes and about the mechanisms by which they have been produced. Further investigations may shed light on the mechanisms underlying the changes; these investigations may involve further examination of the tissue and the application of special techniques, including, for example, histochemical methods. But the process begins with observation.

In looking for changes one should remember that:

1. We tend to see what we are looking for;
2. The likelihood of our finding whatever it is we are looking for depends on our experience of looking for it and finding it.

Unfortunately, what we are looking for might not be seen because:

1. It isn't there, *i.e.*, it does not exist;
2. It has been destroyed during the processing of the material we are examining;
3. It is not visible: a very different thing from not existing.

In addition, we might not recognise whatever it is we are looking for despite it being present. For example, imagine looking for an infiltrate of plasma cells into the connective tissue of the portal tracts of the liver. We should ask ourselves:

1. Do we know what portal tracts are and do we know how to recognise them?
2. Do we know what plasma cells are and do we know how to recognise them?

In both cases the application of special staining techniques will make our task easier. See, for example, Figure 1.1. Having found the portal tracts and having seen plasma cells, we should ask ourselves whether there are more of these cells present than we expected. Perhaps we should count them. We should certainly compare our material with control material in which the number of plasma cells present is assumed to be normal. To take another example: imagine that, whilst examining sections of liver stained with haematoxylin and eosin, you come across well-circumscribed areas that appear paler than the rest of the tissue. What does this mean? Is it an artefact of preparation or a real change that calls for interpretation? Comparison with control sections, or with memory of the normal appearance, will answer these questions.

Our problem is made a little more difficult by the presence of artefacts introduced during the processing of the tissue. We might see

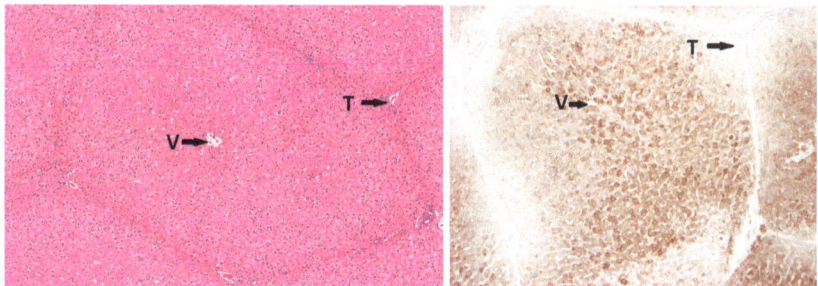

Figure 1.1 **Pig liver.** Standard H&E on the left. Pig liver is particularly good for visualising the lobular pattern. Note the central vein (V) in the middle of the lobule and the portal triads (T) at the apices. H&E cannot, however, demonstrate function. The liver on the right is from the same animal, but this time stained with an immunohistochemical stain for cytochrome 3A4. As you can see, the activity, as demonstrated by the darker brown staining, is far greater closer to the central vein. (Image courtesy of Steve van Crutchen.)

something that looks unusual, perhaps abnormal, and set about interpreting it. But perhaps the abnormality is not real in the sense that it was not present when the tissue was removed from the animal but has appeared since. Cynics sometimes refer to histology as "the interpretation of artefacts". They are half right. Much of what we see is an artefact of processing, but experience tells us which artefacts, or appearances, are important and which are not. Close attention to detail is necessary: we might be looking for changes in the appearance and distribution of chromatin in the nucleus of a cell. Bad preparation might very easily cause changes to the normal appearance; the normal appearance, that is, after good preparation. In the days before the development of the electron microscope mitochondria were studied by the use of special stains and light microscopes. How tissues were fixed was critical to whether mitochondria could be seen: using the wrong fixative during tissue processing would produce entirely misleading results. A last example: Bouin's fixative is an excellent fixative for most tissues and enhances the effectiveness of certain staining techniques. But it is a very poor fixative for the kidney: nobody with experience would try to assess the normality or otherwise of kidney tubules by examination of material fixed in Bouin's fixative.

It will by now be obvious that knowledge of the normal structure of tissues is a prerequisite for the successful practice of histopathology. It will also be obvious that the preparation of the material under study will be critically important. Knowing the appearances of normal tissues after a variety of methods of preparation have been applied is essential. It should also be clear that some familiarity with the use of the microscope is needed.

1.4   CHANGES OFTEN SEEN IN TISSUES

Perhaps the most off-putting feature of histopathology is the vast range of changes in the structure of tissues that have been described. Take any atlas of human or animal histopathology: a quick leaf through will reveal a multitude of descriptions of tumours and other lesions. How could anybody remember all this? The answer, of course, is by experience. Looking at many thousands of sections, recording what is seen, referring to atlases and textbooks: that's how histopathologists acquire expertise. A slow process! Indeed, it takes years to become a competent histopathologist who can identify, classify and sub-classify the tumours of, for example, the ovary. Fortunately, perhaps, most toxicologists need not acquire such experience; they will be interested in the effects of a limited range of compounds or the responses of a limited range of tissues rather than in the whole gamut of histopathology. The toxicological pathologist will need to acquire his or her expertise more slowly.

Histopathological changes can be classified in a number of ways. For example, one might focus on tissue-specific changes. An example of changes only seen in myelinated nerve tissue is shown in Figure 1.2. The joints, the nervous system, the liver: all show characteristic responses to injury. In each case these might involve the acute responses to injury (chemical or physical), and these acute responses are likely to depend in terms of their detail on the tissue involved, as shown in Figure 1.3.

Fortunately for us, these acute responses have similar patterns, although, as just noted, the detail might differ from tissue to tissue. Primary tumours, on the other hand, vary from tissue to tissue: one does not encounter a primary glioma of the liver because there are no glial cells present. Some tumours, however, occur in a variety of tissues: a fibrosarcoma might occur in the liver, in the heart or in skeletal muscle. This is because the cell involved, the fibroblast, occurs in all these tissues. Similarly, a non-malignant response to injury, such as gliosis, is limited to the central nervous system. Malignant tumours metastasize and secondary tumours occur in tissues far away from the primary tumour. The malignant melanoma of the skin metastasizes widely: to bone, liver, lung, brain, *etc.*

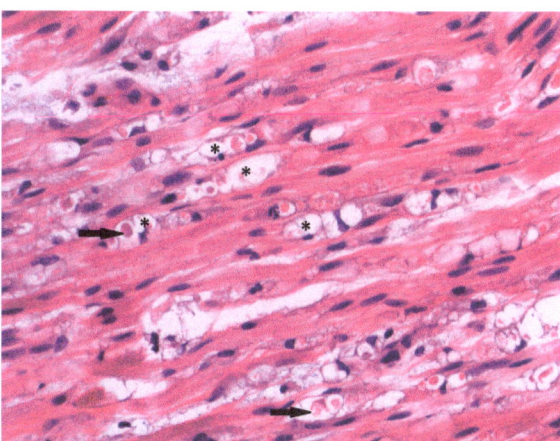

Figure 1.2 **Tissue-specific changes.** Axonal degeneration in the sciatic nerve. Note the digestion chambers represented by the eosinophilic inclusions (arrows) within the clear vacuoles (asterisks). These represent aggregations of denatured myelin.

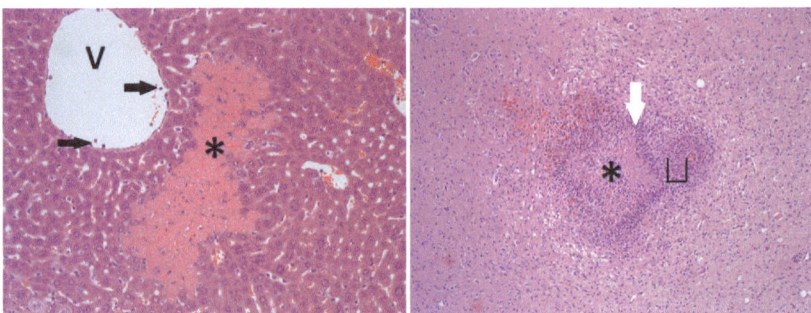

Figure 1.3 **Non-specific tissue changes.** Focal areas of necrosis are common to many tissues. Here, there is an area of focal necrosis (asterisks) in the liver (left panel) and one in the brain (right panel), which is probably the result of a minor infarction. The amorphous eosinophilic appearance is common wherever necrosis appears. In this instance there is a glial response (white arrow) in the brain but, as yet, no inflammatory response in the liver; although, if you look at the central vein (V) to the left, there are a few inflammatory cells beginning to marginate (black arrows). Perivascular cuffing is also visible around the blood vessel in the injured brain (bracket).

The identification of tumours, both primary and secondary, is an important part of the histopathologist's work. In this book we shall not be concerned in any detail with this aspect of histopathology; we will be more interested in the non-carcinogenic effects of chemicals.

On the other hand, one might focus on the process of inflammation or the process of repair and consider how these processes work out in various tissues.

The approach taken in this book is the latter: we shall be interested in how tissues respond to injury. We shall be interested in the similarities between the responses of various tissues and will be touching only lightly on their differences. This seems to me a sound approach: the similarities are more important to the beginner than the differences and understanding the principles is more important than understanding the details, which can be added later. Beginning with the details is always unwise.

1.5   REASONS FOR STUDYING HISTOLOGY

Histology is often regarded, like topographical anatomy, as a descriptive science. Histological techniques may be used in experimental work: histology then becomes an experimental science. Perhaps because of the current tendency to regard hypothesis testing as the *sine qua non* of scientific endeavour, histologists stress that they too are hypothesis testers. Explaining to a grant-giving authority that one is a collector of observations, driven purely by curiosity, with no idea yet of what the collected material might mean and with the avowed intention of passing the material on to someone more gifted with regards to interpretation is not a recipe for success. In fact hardly anyone would advance such a revolutionary view; after all, the age of Darwin collecting fossils and sending them to Richard Owen for study has passed. Libbie Hyman recognised this and wrote in the introduction to her *Comparative Vertebrate Anatomy*: "[speaking of unsolved problems in comparative anatomy and embryology] An army of devoted workers is necessary for elucidating these many questions; but nowadays—alas!—all young biologists want to be experimentalists, and hardly anyone can be found interested in the fields of descriptive embryology and anatomy."[†]

Sir Karl Popper stressed the importance of hypothesis testing; scientists of the distinction of Sir Peter Medawar and Sir John Eccles, both Nobel laureates, have pointed out the importance of Popper's approach. Nobody now doubts the value of the hypothesis-testing approach. What has not been explained is how really clever hypotheses are produced: if this were known then the rate of the advance of science would be much increased. Testing the hypothesis may be difficult but the conceiving of it in the first place is of a different order of difficulty. Of course, trivial hypotheses can be put forward without much effort; putting forward really clever hypotheses is much more difficult and a subject, one might have thought, worthy of the attention of philosophers.

Nobody can conceive an important hypothesis without information. Most of us know that "the ideas come" (if they come) only after much mulling over of our data or of other people's work. The mind needs something to work on before new ideas are produced. This excludes the "flash of brilliance": that, I suppose, sometimes occurs but, I think, largely to those who have thought long and hard about the questions that the "flash" illuminates. Darwin is the classical example of this process. Kekulé is unlikely to have dreamt of his six carbon atoms in a ring unless he had been thinking of organic chemistry, or whatever it was then known as, for some time. It is often said that ideas occur to "the prepared mind": preparing the mind by observation is an important aspect of scientific work.

To those with a biological turn of mind the sequence of collecting, collating, describing, analysing, hypothesising, hypothesis-testing

---

[†]Hyman L H, *Comparative Vertebrate Anatomy*, University of Chicago Press, Chicago, 2nd edn, 1942.

and reporting will be very familiar. Some individuals are very good at the first few steps of this process, others are better at the latter few steps. Because of the attitude often taken with regard to the first few steps of the process, these are sometimes called "hypotheses-generating" studies, the importance of including the word "hypothesis" in the description is apparent to many who seek funding for their work. It is, however, a great error to think that, without a plan to make one's way through the entire process, no useful work can be done. J Z Young's discovery of the giant axons in the squid led on to Hodgkin and Huxley's Nobel Prize winning work on the action potential. Such work was not in Young's mind when he described the giant axon: he produced the description; Hodgkin saw its importance in providing an experimental model for the mammalian axon. It is possible that Hodgkin would not have been interested in describing axons in invertebrates. Young was.

It is often, and entirely incorrectly, assumed that, once a structure has been described, further description is pointless. Thus, the ill-informed regard anatomy and, to some extent, histology as dead subjects. The very fact that they involve description rather than experimentation seems to count against them. Of course this is nonsense. No description is ever complete and there are always new angles from which a structure may be examined and described. These "new angles" include different methods, for example, the use of different microscopic techniques. The electron microscope revealed a wealth of information on structure that had not been visible with the light microscope. Different staining techniques, histochemical methods, autoradiography, phase contrast microscopy, interference microscopy, cell culture… the list of techniques that can be applied to the examination of structure is literally endless as new techniques are being developed continuously. For example, immunohistochemical detection of myelin in the brain can delineate structures that may not be clear using standard histological stains (Figure 1.4). In addition, description is inevitably incomplete because no-one can claim to have examined any structure for all of its life and under all the variations of physiological and environmental factors that affect it. Aging, reproductive cycles, the diurnal cycle, shorter physiological cycles, the influence of environmental factors, disease and death all affect structure whether considered at a macro- or microscopic level. To complete the study of these factors on all types of cell, to choose just one level of analysis, would be the work of thousands of lifetimes.

An aspect of structure that is not sufficiently recognised is the organisation of cells into functional groups. What does this mean?

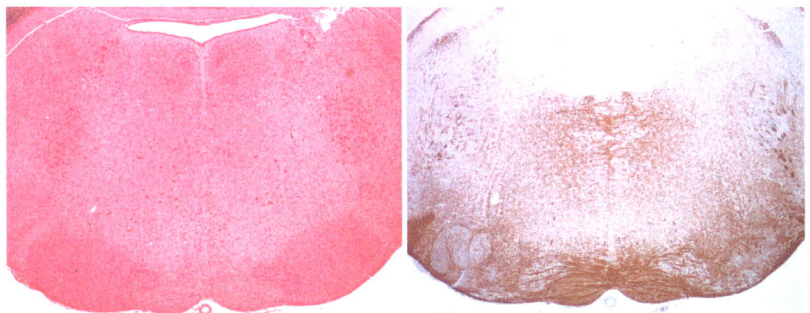

Figure 1.4 **Immunohistochemistry can demonstrate structure as well as function.**
From the H&E image of the pons of the brain (left panel), it is difficult
to see the areas that are myelinated. But when stained using an antibody that detects myelin basic protein, the positively stained brown
areas clearly show the myelinated areas (right panel).

It has become apparent that the functioning of tissues, for example,
skin or liver or the brain, depends on the organisation of cells into
functional groups. This has led to a new way of looking, for example,
at the cerebral cortex: the obvious arrangement in horizontal layers
might not be as important as the less obvious organisation into vertical columns. Much the same point applies to the epidermis. The
liver is organised as lobules: these were described more than one
hundred years ago. The classical hexagonal lobule is, in some species,
a very obvious basis for the structural organisation of the tissue. But is
it a *functionally* important pattern of organisation? Perhaps not: the
Rappaport acinus seems, in fact, to be much more important if not
quite as geometrically satisfying. Pattern of organisation is perhaps
the key to understanding the structure of tissues. The function of a
tissue may well be regarded as an "emergent property" that depends
on the organisation of the cells of the tissue into groups. No understanding of this property will be possible unless these groups are
defined and studied. Such an appreciation leads one to wonder how
many well-described patterns of organisation of cells into tissues are
actually misleading. Consider, for example, the C cells of the thyroid
gland. These cells have often been described as inter-follicular in
distribution. Is this correct or is it an artefact of sectioning? If the C
cells lay amongst the epithelial cells of the follicles but close against
the basement membrane, they might well appear as inter-follicular in
some sections. But if they lie amongst and presumably in contact with
the more ordinary cells of the follicles, what is the significance of
this? Do they interact with the follicular cells? Or are they there "by

chance". Biologists rightly distrust explanations based on "chance". Chance mutation drives evolution, but only the mutations that lead to some advantage survive.

Finding new angles of sight is difficult; this is because our imagination is limited. Widening our imaginations is an important function of education. As pointed out above, nobody seems to know *how* to produce clever ideas. It seems that such ideas are more easily produced when problems of interest to oneself are studied: the reasons for choice of subject or problem are undoubtedly many and varied. Some people know they want to be scientists but take up biology because they are weak mathematicians or because physics seems beyond them. Some take up biology because, wrongly, they see it as an easier subject than physics. Others, usually those with mathematical ability, take up physics because less memory work seems to be involved than in, say, geology or biology. Sir Andrew Huxley could, no doubt, have made advances in many areas of science, but his interest in physiology and bio-physics apparently stemmed from his interest in microscopy. Having found an area of personal interest and appeal it is important to "loosen the mind", to move beyond the descriptions and explanations available in the textbooks and to begin to think. This is so difficult that only a few succeed. One very important reason for this is the weight of what appears to be known: it seems difficult to imagine that anything new can be found. Nothing could be less correct. One need only look back one hundred years or so: the biologists of 1900 presumably thought they were as weighed down as we are with the weight of accumulated knowledge. Experience has shown that some of this accumulated knowledge has turned out to be mistaken or limited or misinterpreted or perhaps not very important and new knowledge has accumulated. Much of this, too, will appear mistaken or limited or misinterpreted or not very important in another hundred years.

Many of those who are drawn to histology tend to be collectors. Collecting, whether collecting facts or pieces of tissue or histological sections or books on histology or on microscopy or on staining techniques, or reprints of published papers, tends to be deprecated by experimental scientists. This is foolishness: Darwin's example as an obsessional collector should be recalled. I suspect that the advance of science has been retarded by this attitude: many natural collectors are put off by such criticism and move away from their subjects or become dull and unimaginative. Collection-based research is seen as old-fashioned and for that reason is criticised. Again, this is foolishness.

Consider, for example, a trainee pathologist who develops an interest in chronic bronchitis. If asked why he had become interested in this, he might say that it is a common and unpleasant disease and that he would like to do something to advance knowledge of the condition. This is not at all an unreasonable position from which to start. He will, of course, know that many others have had similar thoughts and will soon be told that the shelves of the library are groaning under the weight of books and papers on chronic respiratory diseases. He may also be discouraged by observations from his colleagues to the effect that he knows nothing of genomics or molecular biology, and thus is unlikely to acquire a grant for his work. Discouragement is likely and our pathologist might well abandon his project. What a pity this is! Who can say what might have been discovered? Perhaps, even probably, nothing, but how can one know this? If we could predict in advance which projects would produce valuable results or if we could predict which workers would produce such results and which would not, progress would be much advanced; but we cannot.

Let us imagine that our pathologist perseveres and ignores the advice of his apparently cleverer, more successful or more up to date colleagues. He might proceed as follows:

Build up a collection of material: fixed tissue, paraffin blocks and sections from patients dying from chronic bronchitis or from biopsies taken from patients suffering from the disease. It might be argued that no "science" was being done and that this approach seemed rather 19th century in style. Perhaps so, but the activity is far from pointless. Material from different age groups (mainly elderly but not all perhaps), from men and women, from people with different origins, from people treated in different ways, from those that have survived a long period of disease and from those who died more rapidly will build up. Whilst this collection is being developed, our pathologist will be reading the old literature and the more recent. And an idea may occur to him.

Let us imagine, just for fun, that he has noticed that the serous demilunes of the sub-mucosal glands seem to disappear in those who have had chronic bronchitis for many years. He might have simply noticed this on casual examination of many sections. Or he might have noticed this by carefully examining many sections and recording his findings. The latter approach is more likely to turn up a finding of possible interest. Is it an important observation? Has it been made before? Reference to the literature and to the textbooks suggests that it has been noticed but not much has been made of it. If our pathologist is unlucky, a great deluge of papers dealing with the observation will appear from a literature search. This will be discouraging and the conclusion that nothing new is likely to be discovered might be drawn, but also perhaps not. We will now abandon our pathologist and imagine we are conducting the research ourselves. Let us ask some questions.

What is actually known of the serous demilunes of the sub-mucosal glands? If they are similar to the serous demilunes of the mixed salivary glands, they will secrete a serous material that may play a role in flushing the mucous secretion of the cells of the mucous acini from the glands. Or is this far too simplistic: simple flushing!? Do the serous demilunes produce a fluid that is critical in the control of the viscosity of the secretion of the glands? Perhaps they produce enzymes that act on the secretions of the mucous acini and control its viscosity? Perhaps the fluid adds to the mucus and reduces its viscosity? But how does the mixing occur? Perhaps the secretion of the demilunes contains anti-bacterial agents that prevent bacteria becoming established on the airway surface? What can be learnt of this serous material? Histochemical methods will tell us something of the composition of the secretion. How many of these demilunes are there? Or, better, what is the ratio of serous demilunes to mucous acini in the normal gland? Does this ratio change rapidly as chronic bronchitis develops or are the serous demilunes preserved until later in the disease? Can we measure this ratio using image-analysis techniques? Can we express our results in quantitative terms by use of stereological methods? Can we build a three dimensional model of the glands and explore the relationship between the demilunes and the mucous acini in more detail? Are the serous demilunes present in children? Does the ratio of demilunes to mucous acini change with age? Is there a sex-difference in the ratio of demilunes to mucous acini? Do the serous demilunes show a preliminary stage of hypertrophy before they disappear? And are they simply crowded out by hypertrophying mucous acini or are they destroyed by inflammation? Are there signs of mucous metaplasia amongst the cells of the serous demilunes? Do they actually disappear or are they converted into mucous acini? Are the cells of the demilunes apoptotic? Are there basket cells associated with the serous demilunes? If so, do these disappear too? Are the sub-mucosal glands affected by chronic inflammation?

Many questions arise after just a little reflection: there is certainly room for research here and, to return to him, our pathologist should not feel depressed or oppressed. Perhaps most interestingly, whilst following up ideas on serous demilunes, other ideas might occur. Lord Zuckerman pointed out that most of his most successful research began with curiosity about something, but then developed into a study of something quite different. Perhaps the ducts of the submucosal glands or the pattern of innervation of the glands will attract attention and the whole process of asking questions and testing ideas will start again. Of course all this might come to nothing: nobody can know this in advance. But even if it does, not all is lost. A fine collection of material that can be studied by others will have been built up. The Human Tissue Act of 2004 has made this sort of study very much more difficult and it is probably fair to say that pathologists working with human tissue will either not be able to do this sort of work in the future or that they will be required to justify it in so much detail and jump through so many legal hoops that they will choose not to do it. That this is likely to damage pathology and the training of pathologists seems likely. The experimental pathologist is not faced with such difficulties and should archive material enthusiastically.

But none of this would qualify as "experimental research". It is observational research but involves the asking of questions and the proposing of hypotheses and the testing of these hypotheses by further observation: it is "proper science" after all. In all this we are looking for new angles of attack: new questions, the application of new methods and looking with a fresh, but prepared mind.

1.6   ORGANISATION

Organisation is the essential feature of living material. Organisation is both spatial and temporal. Spatial organisation implies the structured distribution of chemicals within the cell. The structural component of spatial organisation of cells is, to a very large extent, membrane. Membrane separates the interior of the cell from its environment. The separation is selective with a regulated two-way traffic of material into and out of the cell. Membrane also defines the nucleus and the subcellular compartments: mitochondria, lysosomes, Golgi apparatus and the endoplasmic reticulum. These internal membranes are also selective. Enzyme systems are segregated by membranes. When segregation is lost, the chemical organisation of the cell falls apart and the cell dies. Chemicals that damage membranes cause a loss of structural organisation of cells and tissues. Chemicals that react with chemical components of cells (for example, chemicals that inhibit enzyme systems) damage the chemical organisation of cells. The form of the cell is controlled by skeletal proteins within the cell. These act as flexible beams that support the membrane and hold it in shape.

Temporal organisation implies the life cycle of the cell and the shorter cycles of activity that may originate within the cell or be prompted by chemicals acting on the cell, for example, hormones or transmitter substances if the cell is under neurological control. Tides of activity ebb and flow through the life of the cell.

Tissues are made up of cells. The cells interact in such a way as to produce the form of the tissue. No part of the tissue is permanent. Components of the cells turn over as do the cells themselves. Extracellular material, including collagen fibres and elastic fibres, are also turned over, being destroyed and replaced. The mechanisms whereby the structural organisation of tissues is maintained are poorly understood. One might think of a blue-print that defines tissue structure. Where is this blue print located and how is the design implemented? Sir W E Le Gros Clark compared the appearance of tissues with that of a reflection seen in running water. He quoted J Needham in saying: "The problem of organization is the central problem of biology...the riddle of form is the fundamental riddle."

More progress has been made with organisation than with form.

1.7   HISTOLOGICAL SECTIONS AS FRAMES FROM A CINE FILM

Cells and tissues are three dimensional: histological sections present a two dimensional image of the tissue or cell. Adding the third dimension is possible by reconstruction from serial sections and, to a lesser extent, by focusing up and down through the thickness of histological sections. Confocal microscopy allows this to be done very effectively: optical sections are observed. But there is also the fourth dimension, so to speak, of time. The histological section might be thought of as a frame from a cine film of the life of the cell or tissue. To ask the histologist to look at one frame and to deduce what has happened before the section was prepared and what would have happened if the tissue or cell had continued to live, is to ask a lot. In fact the problem is more difficult than that. A frame from a cine film shows a picture with considerable depth of focus and interpretable detail: actors might be seen standing in a room or perhaps in a wood or on the deck of a ship. They might be dressed as naval officers or as cowboys. One might be lying on the ground with the others looking anxiously at him. The sun might be setting, or a storm might be blowing. A number of clues to the story are likely to be available. If only the silhouettes of the figures could be seen against a backdrop the problem would be more difficult. Consider how very much more difficult the task would be if a horizontal section through the actors and scenery were to be taken and examined. Now the actors would appear as ovals or blobs and not look like people at all. To deduce the story from such a picture would be impossible. In light microscopy a reasonable number of clues remain: we might hope to see many cells, to be able to look at their spatial orientation and how they are interacting with one another. In electron microscopy (EM) the clues are fewer: without some history, or some appropriately orientating light microscopy, electron microscopy becomes very difficult. Anybody can identify a rat kidney as a kidney, assuming they know roughly what a kidney looks like. Identifying kidney tissue by light microscopy is more testing, but not difficult. But nobody could identify one mitochondrion or a few ribosomes, seen in an EM image, as coming from the kidney. Once the resolution and magnification has increased beyond a certain point, the clues that identify the tissue as a whole are, generally speaking, lost. An example of this can be seen in Figure 4.1 of Chapter 4.

To extend the cine film analogy: the frame is distorted by the processes used in the preparation of the histological section. Thus, we do not see a true frame from the cine film but a frame that has been changed, perhaps in a predictable and fairly well-understood way. Different methods of preparation lead to different distortions of the chosen frame. By comparing sections prepared in different ways, a better picture of what the original frame looked like can be developed even though observation of the true reality is impossible. Examination of living cells is an obvious suggestion. But it is difficult to observe cells that have not been isolated from tissues and the process of isolation might change the appearance and behaviour of the cells. Phase contrast microscopy avoids the preparative processes of conventional histology, but the range of features that can be observed with this method is limited. In histology applied to toxicology it is important to freeze the frame in such a way that damage to cells and tissues is preserved. This adds to the problems: studying a normal cell is difficult, but studying one that has been undergoing a complex series of changes in response to some chemical is more difficult. No cell or tissue can be studied by histological methods more than once in time. Treating a group of animals with some chemical, killing at intervals and studying cells and tissues provides some guide to the time dimension of the changes seen.

1.8   PLANNING TOXICO-HISTOLOGICAL OBSERVATIONS

1.8.1   Defining the Question

As in all research, the first thing to decide is: what specific question are you trying to answer? Unless you are clear about this you will not find it easy to plan your histological study. Even before this question is answered or even asked, you need to be clear about your programme. Toxicological research may be divided into a number of categories and the work may be directed to finding out about:

- A specific effect, for example, pulmonary fibrosis or damage to renal tubules;
- A specific compound or group of compounds, for example, organophosphate insecticides or nano-particles;
- How an organ, such as the liver, responds to a range of toxicological challenges;
- A specific mechanism, for example, carcinogenesis;
- How a specific species responds to a specific compound or, perhaps, to a range of compounds, for example, how fish respond to effluents discharged into rivers and lakes.

Of course, these categories may be combined: you may wish to find out about how the human lung responds to fibres and the focus of your study may be on the fibrotic response of the lung. Defining the focus of interest will help to define the methods that should be used. If you are a fish or insect toxicologist, you will, rather obviously, need to apply methods appropriate to these animals. If you are interested, per se, in renal tubules, then you may wish to study their responses to a range of compounds.

Thinking in terms of the following is helpful:

- What is the question we are trying to answer?
- What will we be looking for in our histological study?
- Which histological method or methods are best suited to revealing what we are looking for?
- How may the methods be refined to yield optimal results?

For example, imagine that we are interested in a compound, C, that is chemically related to other compounds that have been shown to cause pulmonary fibrosis in man when inhaled. Reasonably enough you will be thinking of an inhalation study or a study involving instillation of compound C into the lungs of some animal model.

It would be eccentric to suggest that the tortoise should be used as the animal model! You will be likely to choose the rat or the mouse and to argue that these animals have long been used to predict effects in the human lung. You will wish to know whether exposure to C causes fibrosis; therefore, you will need histological methods that allow fibrosis to be detected. You will be interested in the relationship between dose and effect, and thus will want to know about low-level changes: early fibrotic changes. You will also be interested in the time scale of the effect: when, after exposure, do the earliest changes occur? You discover, by reading up on chemicals related to C, that they produce an inflammatory response in the lung and that this is followed by fibrosis. You know that inflammation is associated with leakage from capillaries and the deposition of fibrin in intercellular spaces, which may be enlarged by the leakage of fluid. Differentiating between fibrin and collagen, the basis of fibrosis, will therefore be important to you and you will need to select a histological method that will allow this.

A little reflection, as the geometry textbooks used to say, will show that haematoxylin and eosin staining would be a poor choice. Of course you might feel that staining with H and E would provide a useful over-sight: you would be correct, but it won't tell you much about fibrosis. Choosing a staining technique such as Lendrum's MSB or Masson's trichrome would be sensible. You might have thought of the van Gieson method: this is not reliable for "young collagen" and you will need a method that is reliable because you are interested in identifying early changes. You might recall that reticulin fibres, or fibres that stain like reticulin fibres, often appear before collagen appears: a silver impregnation method might be attractive.

We have jumped ahead to thinking about staining. We should, of course, have thought about fixation first. How will you fix the lungs of your rats or mice? By removing the lungs and dropping them into a beaker of fixative? By slicing the lungs into 5 mm slices and dropping those into a beaker of fixative? By expanding the lungs with fixative, tying off the trachea and then dropping the lungs into a beaker of fixative? The latter is by far the best method when looking for fibrosis. Indeed, it is always the best method for fixing lung unless you are looking for pulmonary oedema: the oedema fluid is diluted by the fixative and becomes more difficult to identify in the stained sections. And which fixative will you choose? You might stick to formalin, preferably buffered, or you might recall that the collagen stains are enhanced by fixatives containing mercuric chloride and choose Zenker's fixative. This would be a rather good choice. If you intend to use silver impregnation to identify reticulin fibres, for example, Gordon Sweet's method, then formalin might be a better choice for a fixative, but the choice of fixative is not critical.

Paraffin sections will be perfectly acceptable: there is no reason, at least no reason so far, for preferring celloidin or frozen sections. Paraffin sections of lung may be cut a little thicker than paraffin sections of other, more solid, organs; 8 or 10 μm sections would be appropriate. Even thicker sections might be used, but if the fibrosis is well-developed the staining might well be a little too heavy to allow critical examination.

We have argued our way to a plan: lungs fixed by expansion with Zenker's fixative, paraffin sections cut at 8 μm, stained with: H and E (because that's traditional!), Lendrum's MSB, Masson's trichrome and Gordon Sweet's reticulin methods. A nice set of stains! If you are following up published studies on compound C, then checking to see what methods others have used would be sensible. Slavishly following other people's methods is, however, an error.

Similar thinking might be applied to the following questions:

Does C cause:

Pulmonary oedema, centrilobular or periportal necrosis of the liver, aplasia of bone marrow, damage to germinal centres of lymph nodes, damage to beta cells of the pancreatic islets?

The point is that, in each case, you should reason your way to an appropriate set of histological techniques. The techniques required will differ as the effect varies.

1.8.2   Choosing Staining Methods

For overall appearance and tissue architecture, H and E is satisfactory. This standard method will allow you to identify major changes, such as the appearance of tumours in the midst of essentially normal tissue.

For changes in connective tissue (for example, fibrosis due to deposition of excess collagen or elastosis, which means destruction of elastic fibres), a connective tissue stain is needed. Different stains are needed for collagen fibres, elastic fibres and reticulin fibres.

For other features, for example, mucin or fat, appropriate stains and, in the case of fat, processing methods should be selected.

If changes in the appearance of nuclei are likely to be critical features, then Weigert's iron haematoxylin would be an excellent choice.

See Chapter 8 for details of these methods.

1.8.3 Planning the Observations

It seems obvious but it is worth saying that histological observations depend on being able to distinguish the abnormal from the normal. Thus, controls are always done and you will have access to normal tissue. This should, of course, be processed in exactly the same way as tissue from the animals exposed to C. What you are learning from the controls is the normal appearance of whatever tissue you are looking at **after processing, including staining, with the methods you have chosen and, at least as importantly, as these techniques work** *in your hands*. Textbook pictures of the normal appearances of tissues are no substitute for your own controls.

1.8.4 Refining the Methods

One of the drawbacks of limited space afforded to the description of "methods" in scientific papers is that histological methods are usually dismissed in a line or so: "sections were stained with haematoxylin and eosin or with Masson's trichrome." Some staining methods are comparatively automatic and do not require more than average care when applied; others are much more difficult to use well. Methods that require differentiation to bring out certain features, for example, collagen fibres and to allow then to be clearly distinguished from, say, muscle, require care. The control sections may not help as much as one would wish: the changes sought in the sections from animals exposed to compound C will not be present in the controls. Consider, for example, the development of fibrosis. This often follows inflammation and, during the inflammatory phase, one might expect fibrin to be deposited in interstitial spaces. Distinguishing fibrin from collagen will be necessary. Necessary, but not easy: with some methods the staining of fibrin varies with the age of the deposits; early on it may stain clearly as fibrin but , later, it may stain as do collagen fibres. I have seen fibrin deposits in pulmonary alveoli stain well with Lendrum's MSB method, the fibrin is revealed as a tangled mass of red-staining threads, but in the centre of the mass of fibrin pale blue amorphous deposits were seen. Was this collagen? Collagen certainly stains blue with Lendrum's MSB, but it is difficult to believe that collagen was being laid down in what was, essentially, a clot. The fact that the blue-staining material was amorphous in nature and not fibrous persuaded me that it was unlikely to be collagen. Or perhaps the method used for fixing the tissue, Zenker's fixative for the usual period, was not adequate: Lendrum recommended long fixation in formal-sublimate (up to eight weeks!) before his MB stain was used.

Perhaps van Gieson's stain would have been a better choice: red collagen *versus* yellow fibrin. The disadvantage of this method is that young collagen does not stain as deeply as mature collagen. It is very easy to wash out the red acid fuchsin during dehydration of the sections and losing the staining of young collagen is easy. Masson's trichrome differentiates fibrin from collagen (red *versus* blue or green depending on the stain used) and it is widely used. The tricky step with this method is differentiating the red cytoplasmic stain, ponceau 2R and acid fuchsin, with phosphomolybdic acid so as to leave the collagen unstained and ready to take up the blue or green collagen stain. It would not be difficult to remove the red stain from fibrin during the differentiation step.

The message is that care will be needed.

1.8.5   Identifying the Changes

Chapter 4 deals with how to examine histological sections: the advice given there will not be previewed here, only a few points will be mentioned. The following examples include a lot of detail that you might not, at present, understand. Much of this will be explained in subsequent chapters.

The expert can tell "at a glance" whether a section is normal or abnormal: an enviable skill that only comes with experience. It is reassuring to recall that every expert was not an expert when he or she began! So how do you start? Because it is always easier to think about a real example than about abstractions, let us take an example.

Example 1

Objective: To study the effects of a compound suspected of damaging the conducting airways of the lung.

Question: Does compound C, known to be chemically related to, or in some way similar to, compounds known to produce damage to the conducting airways of the lung, damage the conducting airways of the lung and, if so, how dangerous is it?

We know, for example, that high concentrations of sulphur dioxide damage the conducting airways and we may ask, perhaps, whether nitrogen dioxide or hydrogen chloride have the same or similar effects. We suspect that they might on the grounds that these compounds dissolve in water to form acids, akin to sulphur dioxide.

Experiment: A series of exposures of rats to the compound of concern are undertaken; exposure at varying concentrations and for varying times; animals are allowed to recover and are killed at intervals up to a few weeks after exposure. Controls, treated in precisely the same way, barring exposure, are also provided.

Planning the histological observations: Our focus is on the effects on the conducting airways. We need a method of fixation that will fix these airways effectively: slicing the lung and placing the slices in fixative is likely to be suitable. If we were concerned with the alveoli we might prefer to distend the lungs with fixative, but we might be concerned that this would displace an inflammatory exudate from the epithelial surfaces of the conducting airways and push it down into the alveoli. We settle for fixation of slices of lung. The fixative? Well, formalin (buffered formalin: see Chapter 5) would be a standard choice. Looking ahead, we shall not be much interested in collagen fibres (the response being studied is an inflammatory response, although, of course, it might be followed by some fibrosis) and so there should not be a special need to enhance the staining of collagen with aniline dyes by fixing with a mixture containing mercuric chloride. Blocks are taken and processed in the usual way. Rat lungs are small and whole coronal sections can easily be processed: no need, therefore, to worry about sampling from the central regions or the periphery of the lung. The sections are stained with haematoxylin and eosin: a rather unexciting choice but one likely to be made. A much better choice would be Masson's (or Mallory's) trichrome, even with formalin-fixed tissue. Sections are stained and mounted. (If worried about the possibility that a trichrome collagen stain will be called for and you have doubts about the formalin fixation: take heart! The sections of formalin-tissue can be mordanted in alcoholic picric acid and mercuric chloride; this will improve the trichrome staining. If Mallory's trichrome is to be used, then post-chroming is helpful: see Chapter 8.)

The examination of the sections: The first question is: how familiar are you with the structures that you will be examining? Everybody starts by being entirely unfamiliar with histological appearances of tissues and you might be in that position. Your reading of the literature relating to the compound you are studying will, of course, have told you a good deal. Reading or re-reading a standard account of the structure of the conducting airways of the rat is a sensible step. Note that you need to be familiar with the structure in the rat, not in man. That requirement might seem to exclude the classic textbooks of human histology. However, reading the accounts provided in these textbooks (see the Bibliography) will give you a background against which to make your observations: you will rapidly notice the difference between rat-conducting airways and human-conducting airways. It is important to note that individual animals will vary even within the same treatment groups so it is important to be well-versed in the variations that can occur within normal histology, as well as any pathological changes that may not be associated with treatment, *i.e.*, infections, *etc*.

As always the section should be scanned at low power. Large airways and their accompanying branches of the pulmonary artery and tributaries of the pulmonary vein will been seen near the hilum of the lung. Only the largest airways have any cartilage in their walls, and sub-mucosal glands are few and far between in the rat. Whilst examining the largest airways, turn to the accompanying pulmonary artery. Look at the muscle in its wall: you might be surprised to see that it is cardiac muscle: striated, with intercalated discs that are clearly visible. Being able to see the histological detail of the muscle fibres is a good test of your stain. An iron haematoxylin stain would bring out these details particularly clearly. Move to higher power and examine the epithelium lining of the large airways. The lumen of the airway should be empty, but desquamated epithelial cells, inflammatory cells and debris may be present. Had the lung been fixed by expansion with fixative, this material might have been displaced into the smaller airways. The epithelium is a ciliated *pseudo*-stratified columnar epithelium: goblet cells are present. The rat relies more on its goblet cells than on sub-mucosal glands for production of airway mucus; the reverse is the case in man.

Does the epithelium look normal? Can you see the layer of nuclei of the basal cells close to the basement membrane and the layer(s) of nuclei of the taller cells nearer to the luminal surface? Is the epithelium complete or is there evidence of cell loss: desquamation? Is the epithelium infiltrated with inflammatory cells? It is quite possible to see cilia at high power but they seldom look quite as perfectly regular as shown in pictures in the textbooks. The rat respiratory epithelium contains a fair number of unciliated cells; these include Clara cells. H and E is not an ideal stain for examining cytological details: Heidenhain's iron haematoxylin is a much better stain for this.

Now examine the basement membrane and the sub-mucosal tissues. If a brisk inflammatory response was underway when the animal was killed, infiltration by inflammatory cells will be obvious. Capillaries may stand out clearly: their lumens (or lumina, but not lumena if you did Latin at school) will be filled with red cells.

The sub-mucosal tissue may look oedematous: pink staining spaces between the cells may be seen. It is unlikely that the smooth muscle of the wall will be much affected and the cartilaginous plaques of the wall (if there are any in the wall of the airway you are examining) will appear normal.

If you have seen evidence of inflammation then grading this is the next step. How severe is the damage? The control sections will provide a benchmark of normality; the high-dose animals will provide a benchmark for the other end of the scale of damage. Examining selected sections from groups exposed to different doses will be helpful in defining a spectrum of damage. It is likely that the scale or spectrum of damage will show some step changes. For example, loss of epithelial cells is unlikely at low doses but will appear as doses increase. Similarly, basal cells are likely to remain attached to the basement membrane at higher doses than other epithelial cells. Defining the doses at which basal cells begin to be detached might be important. Defining minimal levels of injury is never easy. Randomising the sections from the controls and from the exposed animals can help with this. Examining the sections blind and grading the normality, or rather the abnormality, of the tissue is the best way to define the dose that produces the lowest level of injury that you can recognise. You may feel that an expert could do better: no doubt this might be true but, if you adopt a strategy for examination and note the appearance of features in turn, you will do very well.

You may notice changes in the sections, which could be illuminated by special stains. For example, if you think that goblet cells are being emptied or lost from the epithelium, then staining for mucus is an obvious idea. Mucous cells that have discharged their mucus may retain a few droplets in the "stem of the goblet" and it is here that new mucus appears as the cells recharge. You will not detect this with H and E but periodic acid Schiff (PAS) staining will make it clear. It is known that some irritants, cigarette smoke is an example, has differential effects on mucous cells producing neutral or acidic mucus: staining with Alcian Blue at different pH levels will bring this out. Always cut and mount more sections than you need; this will allow you to go back and use special stains. Using a special stain on the next section after the last stained with H and E is especially instructive.

Example 2

Objective: To study the effects of a compound suspected of causing damage to the proximal tubule of the kidney.

Background: Let us imagine that the compound being studied is related to others that are known to damage the proximal convoluted tubule (PCT) of the kidney. Knowing something about the PCT will clearly be necessary before you look at the sections. The PCT is perhaps the busiest part of the renal tubule that, with the glomerulus, makes up the individual nephron. About 85% of the fluid filtered at the glomerulus is reabsorbed here: the reabsorption is isotonic in that sodium chloride is reabsorbed with the water and the fluid leaving the PCT has the same osmotic pressure as the fluid entering the PCT. Glucose and other low molecular weight molecules are transported across the PCT to the blood. Small protein molecules, which had passed through the glomerular filter, are also reabsorbed by vesicles. Professor D B Moffat (Department of Anatomy, Cardiff University) used to joke that the kidney worked on the principle of throwing the baby out with the bathwater and then recovering the baby. The PCT does most of the recovering. Fixing the epithelium of the PCT so as to reveal its *in vivo* appearance is notoriously difficult: in "ordinary" histological sections it often looks far from normal. We shall consider fixation before dealing with staining and the normal appearance of the PCT.

Fixation and processing: The PCT rapidly deteriorates, *via* autolysis, as soon as the animal is dead. The kidney is best fixed by intra-vascular perfusion: warm normal saline (0.9%) is run through the circulatory system to flush out blood, followed by fixative. Tubules near the surface of the kidney can be fixed by dripping fixative onto the surface of the living kidney: the kidney being exposed by opening the abdomen of an anaesthetised animal. Neither method is of much use in routine toxicological work; the usual approach is to remove the kidney as soon as possible after death, slice it so as to reveal a classic broad-bean-shaped surface and drop the slices into fixative. Formalin is a satisfactory fixative. The only standard fixative that is thoroughly unsatisfactory, for some unknown reason, is Bouin's fixative. Cooling the fixative will delay autolytic changes in the kidney. Doing an experiment to compare the effects of fixative at room temperature and at 4 °C, before starting on the main study, would be sensible. Cooling retards diffusion of fixative and this needs to be balanced against the advantages of a delay in autolysis. The slices are processed whole: they will not be too large to fit in a cassette. This assumes you are working with rats or other small mammals! Embedding in paraffin is the routine method but semi-thin plastic sections would be very much better for assessing tubular cells.

Staining: H and E, as ever, is the routine staining method. It is perfectly satisfactory for finding your way about the histological topography of the kidney, but PAS or methenamine silver staining for the basement membrane is often helpful. If you are really interested in cytological detail then Heidenhain's iron haematoxylin is the best method.

Examining the sections: The first step is to review the histology of the rat kidney to get your bearings, so to speak. Everybody knows that the kidney can be divided into a cortex, containing the glomeruli and part of the tubular system, which is arranged around the medulla that also contains parts of the tubular system but no glomeruli. The medulla projects as a papilla into the pelvis of the kidney. Urine flows from the collecting ducts of the medulla into the pelvis (pelvis: basin) and into the ureter. The cortex and medulla can be distinguished with a magnifying glass. The glomeruli are easy to identify at low power: they appear as small reddish dots in the cortex. Closer examination of the medulla shows that it can be subdivided into an inner and outer region, and that the outer region can itself be divided into an inner and outer stripe.

Examine a dozen or so glomeruli at high power (oil immersion is not necessary at this stage). Note that the glomerulus contains a knot of capillaries. The capillary knot is connected to afferent and efferent (going in, coming out) arterioles, which can be seen in cross section near the glomeruli. Look for a glomerulus in which you can see an arteriole connecting with the capillary knot. This defines the vascular pole of the glomerulus. You will not see this at every glomerulus: the randomness of the plane of sectioning of the glomeruli will defeat you. The other pole is the urinary pole, where the cavity of the glomerulus is continuous with the PCT. You will have no difficulty in seeing cross sections of tubules around the glomeruli. The cross sections are round or oval and the tubules are lined with a cuboidal epithelium. Even without looking too hard, you will notice that the cells of some tubules are rather pinker than those of others: the ones with the rather pinker staining cells are the PCTs. The others, the ones with the less pink cells, are sections of distal convoluted tubules (DCTs) and collecting ducts (CDs). Being able to distinguish the PCT from the DCT and the CD is important. The Table 1.1 will help you to do this.

Necrosis of the PCT: The earliest change is that of cloudy swelling, a singularly difficult change to detect in poorly preserved epithelium! As the damage progresses cells are lost into the lumen of the tubule.

Table 1.1 Histological features of renal tubules.

Feature	PCT	DCT	CD
Quality of fixation, preservation of epithelium	Often poor	Good	Good
Eosinophilia of cuboidal epithelium	Marked	Bluish-pink, often more bluish than pink	Bluish pink, often more bluish than pink
Nuclei	Few seen and those that are seen appear rather irregularly amongst the pinkish cytoplasm. The cells are large and transections often do not include the nucleus	Regularly arranged nuclei, often six or more per cross section. Nuclei take up a lot of room in the cells and bulge the luminal surface	Regularly arranged nuclei, often eight or more per cross section.
Lateral boundaries of cells	None seen due to interdigitation of the lateral surfaces of the cells	Fairly clearly marked	Clearly marked
Brush border	Present but not well defined in poorly preserved material	Not present	Not present
Basal striations	Present: short, pale purple lines running at right angles to the base of the cells	Not present	Not present
Presence of a well-defined lumen	Often difficult to define the lumen: cells look ragged and sometimes seem to fill all the space within the cross section	Clearly defined	Clearly defined

Nuclei are lost and just one or two, hanging on, so to speak, with a thin rim of pink staining cytoplasm may be all that remains of the epithelium. Depending on dose and on the period since the final dose, recovery occurs: the epithelium regenerates from the remaining cells and mitoses appear. Mitoses are very rare in normal PCT epithelium. Necrosis may be associated with an inflammatory response around the tubules.

The necrosis may be graded from minimal to complete destruction of the epithelium.

It is, of course, a mistake to look only at that part of the nephron where you expect damage to occur. Examine all the nephron. Calcification of tubular cells in the medulla is often found in older rats: deep blue staining of the calcified areas is seen in sections stained with H and E. Special stains, for example, the von Kossa method, will allow the calcium salts to be identified with precision.

1.8.6 Where Next?

Much of what has been said in dealing with these examples will have seemed singularly difficult to those with no background in histology. If you fall into this group, take heart! The following chapters are designed to help you.

CHAPTER 2

An Introduction to Histopathology

Histopathology is the study of the responses or reactions of cells and tissues to disturbance by disease or insult, be it physical or chemical. Toxicologically active substances can damage cells and tissues, and this damage can be studied histologically. Histopathology can be divided into "general", dealing with the responses shown by all cells and tissues to damage, and "special", where the focus is the specific changes seen in individual tissues. In this chapter we focus on general histopathology and consider responses, including those produced or shown by cells when they first respond to injury. Death of cells is also considered, as is the complicated subject of inflammation. Of course, all these responses can be described and analysed in far more detail than has been possible here: this chapter is an introduction to a large subject, but is very important if more complex changes are to be understood.

2.1 THE RESPONSES OF TISSUES TO INJURY

Cells may be damaged by chemicals in a wide variety of ways. One of the most important of these concerns restriction of the cell's capacity to generate adenosine triphosphate (ATP), the currency of energy exchanges within many cells. Such damage may be caused by a reduction in the supply of oxygen or by a reduction in the capacity of the cell to make use of such oxygen as is available. Chemicals that bind to enzymes of the pathways involved in the metabolism of glucose have

Histological Techniques: An Introduction for Beginners in Toxicology
By Robert Maynard, Noel Downes and Brenda Finney
© Maynard, Downes and Finney 2020
Published by the Royal Society of Chemistry, www.rsc.org

similar effects: the energy supply is reduced or perhaps cut off. No cell can survive such an effect for long, although their capacity to do so varies from tissue to tissue depending on what alternative pathways for the production of ATP are available. As the energy supply is reduced, those functions of the cell that depend on ATP are also reduced. One of these is the ATPase-dependent sodium/potassium pump of the cell membrane. If the pump fails, then cells will swell. Swelling of cells and a change in the appearance of the cytoplasm are amongst the first signs of cellular damage.

This first stage of damage is referred to as cloudy swelling. This is a splendid old histopathological term that describes the appearance of the somewhat damaged cell. It does not tell us anything about the mechanisms involved; it merely tells us that the cell is in trouble. Cloudy swelling is seen as a result of damage in, for example, the cells of the renal tubules, the cells of the liver and the cells of the heart. In each case the changes are similar; an example is shown in Figure 2.1. The cells look swollen and the normal outline of the cell is changed. In the liver the hepatocytes look more rounded than usual. Such a change can only be appreciated by comparing a normal section with one of liver that has been slightly damaged. The cytoplasm looks granular: another change that can only be appreciated by comparison with the normal appearance. The cytoplasm remains eosinophilic; in fact it may contain small eosinophilic granules or vesicles.

As the cell accumulates more water, unstained spaces appear in the cytoplasm. When these are prominent, the appearance is described as hydropic degeneration. Care is now needed. Clear spaces in the cytoplasm of cells can be produced by dissolving out of lipids during tissue processing. Deciding whether these clear spaces are due to the accumulation of water of whether they represent a loss of lipid requires the use of special stains for lipids and these will, of course, not work if the lipids have disappeared during processing of the tissue. Thus, frozen sections of fixed material, stained for lipids, will be needed to make certain the diagnosis of the change.

The accumulation of fat is also commonly seen in damaged cells. Fatty degeneration is the term used to describe this and is shown in the lower panel of Figure 2.1. It is seen, for example again, in the liver, in the renal tubules and in the heart. Precisely why this change occurs is not easy to say: increased production of fat, unmasking of fat already present and failure to export from the cell the fat that is produced could all explain the change. Fatty degeneration of the liver is frequently seen after administration of liver-damaging chemicals, such as chloroform and carbon tetrachloride.

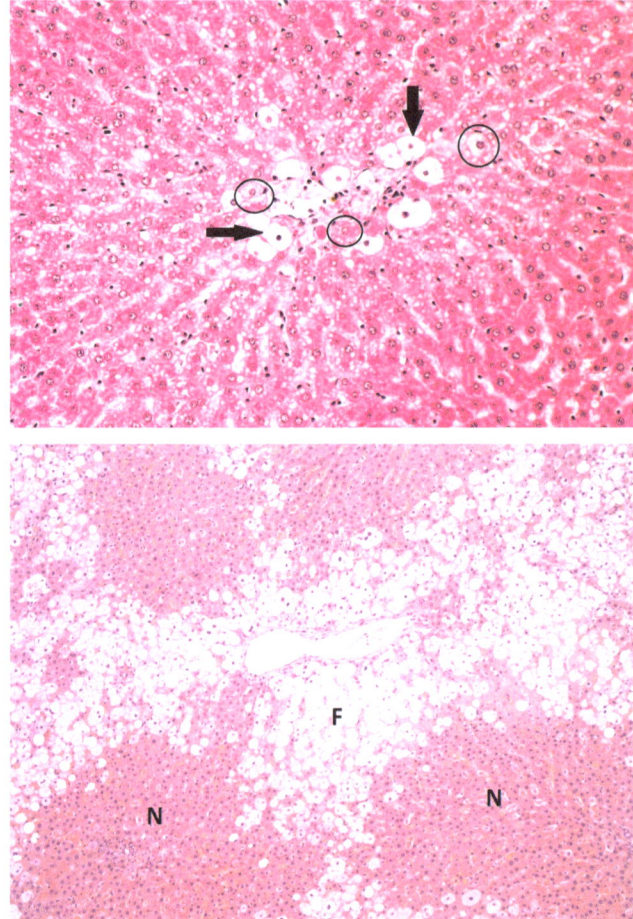

Figure 2.1   **Tissue responses to injury.** Cloudy swelling in the liver 24 h after carbon
tetrachloride exposure (top panel). The early stages of degeneration are
hallmarked by a cellular shape change from hexagonal to rounded
(within circles). Even at this stage, many of the affected cells have gone
a little beyond the stage of early cloudy swelling into hydropic degeneration (arrows). Fatty degeneration in the liver (F, bottom panel)
is seen in between islands of normal liver tissue (N).

Cloudy swelling and fatty degeneration are not irreversible changes.
If the toxicant is withdrawn or the oxygen supply restored, the cells
may recover their normal appearance. However, as damage progresses, irreversible changes appear. These affect both the nucleus of
the cell and the cytoplasm.

Nuclear changes include swelling of the nucleus and a decrease in
its basophilia. The normal basophilia of the nucleus is due to its

content of nucleic acids and, as these break down under attack from chemicals (karyolysis), the density of the staining decreases. Chromatin may also become clumped or scattered as fine granules (karyorrhexis). As a last stage of degeneration, the nuclei become small with very dark staining: this is pyknosis. These changes can only be appreciated when the appearance of damaged cells is compared with that of normal cells, as seen in Figures 2.2 and 2.3.

As these changes occur in the nucleus, the cytoplasm of the cells may become more eosinophilic. We should really say acidophilic because the intensity of staining by an acidophilic dye is increased. Eosin is an acidophilic dye and it is so widely used that eosinophilia

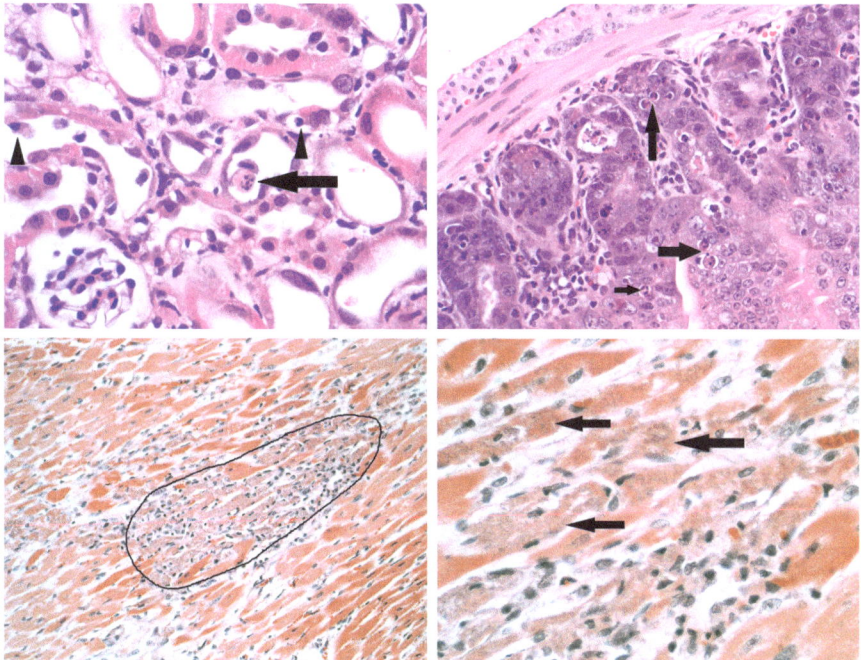

Figure 2.2 **Nuclear changes and calcification.** Apoptosis and karyorrhexis (top panel). Karyorrhexis is not seen all that commonly; on the left panel there is a karyorrhectic cell masquerading as a snowman and wearing a pyknotic hat (arrow). There are several other pyknotic nuclei in the field (arrowheads). The right panel shows a cluster of apoptotic cells in a duodenum of a mouse treated with an anti-cancer agent designed to disturb normal cell cycling (arrows). Calcification of cardiac muscle (bottom panel). At low power magnification (left panel), all that can be seen is a slight tinctorial change and an increased cellularity (outlined). At a higher power (right panel), the basophilic granules that represent early calcium deposition can be seen (arrows).

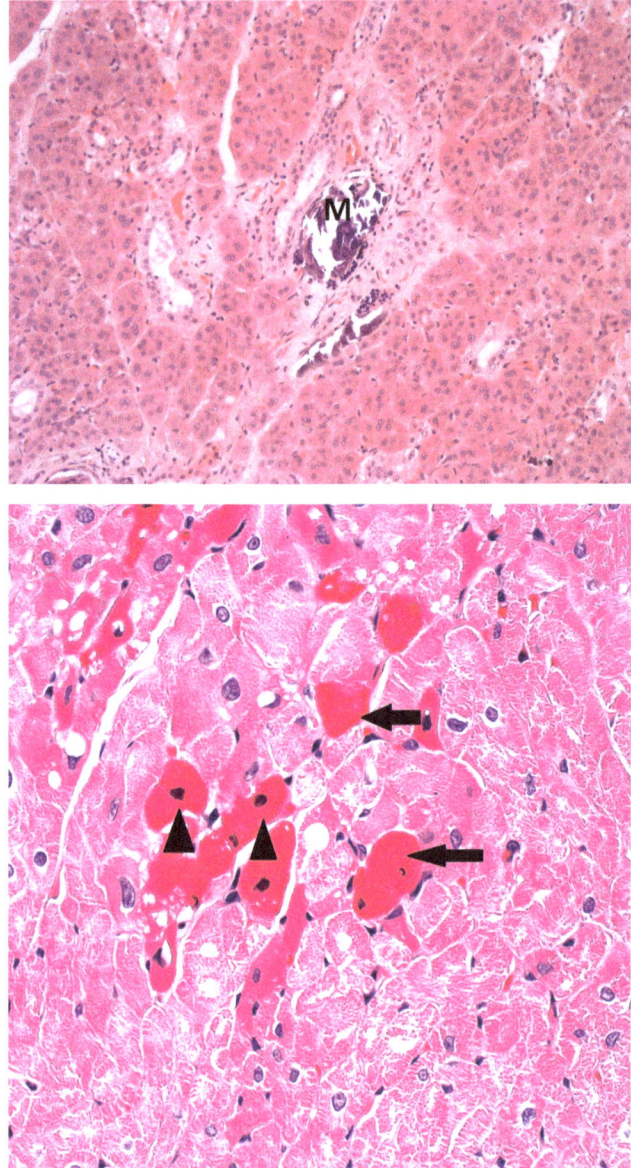

Figure 2.3   **Mineralisation and necrosis.** Vascular mineralisation in the salivary gland (top panel). At this later stage the more familiar purple colour, with its typical crystalline structure, is apparent (M). Acute myocardial necrosis (bottom panel). Note the eosinophilia of the cardiac myocytes (arrows) and the pyknotic nuclei (arrowheads). At this stage, there has not been time for an inflammatory response to develop.

and acidophilia are often regarded as synonymous. This increased staining with eosin can progress to hyalinization: hyaline material is glassy and homogeneous in appearance. Hyalinization affects not only the cytoplasm of cells but also connective tissue: old scars often present hyalinized areas.

Calcium is deposited in damaged cells; the early stages of this are shown in Figure 2.2 and the later stage is shown in Figure 2.3. This is a change that is seen in the later stages and affects, for example, the cells of the renal tubules. Special stains reveal the presence of the calcium. Such deposition of calcium is very common in the kidneys of old rats: chronic inflammatory disease of the kidney explains such effects.

As the cell breaks down or undergoes necrosis, pieces of cytoplasm may be lost. The cells of the renal tubule, for example, take on a tattered appearance: some cells have all their cytoplasm in place, other about half and some almost none at all. These cells are dying and pieces are breaking off into the lumen of the tubules. Necrosis is the final stage: the cells are irreversibly damaged and dying; examples of this are shown in Figure 2.3. As they breakdown and indeed before that, they release inflammatory mediators that call into play an acute inflammatory reaction. Inflammation is a two-edged weapon: further damage may be produced whilst dead cells are being cleared away prior to repair processes beginning to work. We shall discuss acute inflammation in the next section. An acute inflammatory reaction may be succeeded by a chronic inflammatory reaction: the tissue fails to resolve the insult and return to normal. The acute inflammatory response subsides leaving a different form of cellular response (mono-nuclear cells rather than polymorphs) and fibrosis occurs: collagen is deposited. Chronic inflammatory responses are classically seen in diseases such as tuberculosis and sarcoidosis.

Some chemicals induce not necrosis but apoptosis. Both processes involve the death of cells but necrosis is "noisy" and apoptosis is "silent" in terms of signals sent out from the affected cell. Apoptosis is an essential feature of development and occurs without calling out an inflammatory response. Apoptotic cells shrink in upon their nuclei, which become pyknotic in appearance, as shown in a section of duodenum in Figure 2.2.

The changes described very briefly above are the hallmarks of tissue damage. Other changes also occur in disease states. These include amyloid degeneration, gelatinous degeneration and mucoid degeneration. We shall not discuss these changes here: details may be found in the textbooks of pathology listed in the Bibliography. Glycogen may accumulate in damaged hepatocytes and may appear not only within the cytoplasm but also in the nucleus of the cells.

2.2 ACUTE INFLAMMATION

Acute inflammation is the normal response of tissues to injury. All kinds of injury that lead to damage to cells trigger an inflammatory response. Indeed, it is largely the damaged cell that, so to speak, calls for help in dealing with the cause of its injury. Thus, we should not be surprised to find that infection calls out a brisk inflammatory response.

Acute inflammation has long been recognised by clinicians: Celsus in the first century AD described the four cardinal signs of inflammation: heat, redness, pain and swelling. Galen, in the second century AD, added to these: loss of function. Histological techniques do not allow pain, heat and loss of function to be recognised, but redness due to the dilatation of blood vessels and swelling due to the leak of fluid from blood vessels involved in the inflammatory response can easily be confirmed. In addition, histological examination reveals the emigration of polymorphs (white blood cells, largely neutrophils) from capillaries into the area of inflammation. Later, macrophages appear. The entire response is orchestrated by inflammatory mediators released by damaged cells and also by the cells that arrive in the area of damage. A very large number of mediators have been described: we shall not be concerned with these here; our interest will be in what can be seen on histological examination.

Necrosis always causes an inflammatory response; apoptosis does not. The acute inflammatory response is often described in terms of the vascular component of the response and the cellular components of the response.

2.2.1 Vascular Component

The vascular response to injury is focused on the capillaries. Several changes occur: the capillaries become dilated, the flow of blood is slowed, the endothelium of the capillaries becomes leaky, perhaps due to a loosening of the inter-cellular junctions, fluid leaks into the surrounding tissue and, with the fluid, plasma proteins, including fibrinogen, also move out of the vessels. These changes account for the redness and swelling described by Celsus. The leak of protein-rich fluid is an important feature of the acute inflammatory response. Fibrinogen exposed to the debris of cellular damage is converted to fibrin and the intercellular space becomes filled with a sticky fluid. This is important in limiting the spread of bacteria and raises an interesting point. The acute inflammatory response is a generic response to tissue injury: the same response is produced whether the cells are being damaged by bacteria or by chemicals. Thus, we should not be surprised when we fail to think of a "purpose" for certain features of the response to injury by chemicals; it is likely that the response evolved largely to deal with infection.

Dilated capillaries full of red cells are a characteristic feature of histological sections taken from damaged tissue. If damage to a surface epithelium has occurred, the capillary response will be seen in the connective tissue just below the epithelium, as shown in Figure 2.4.

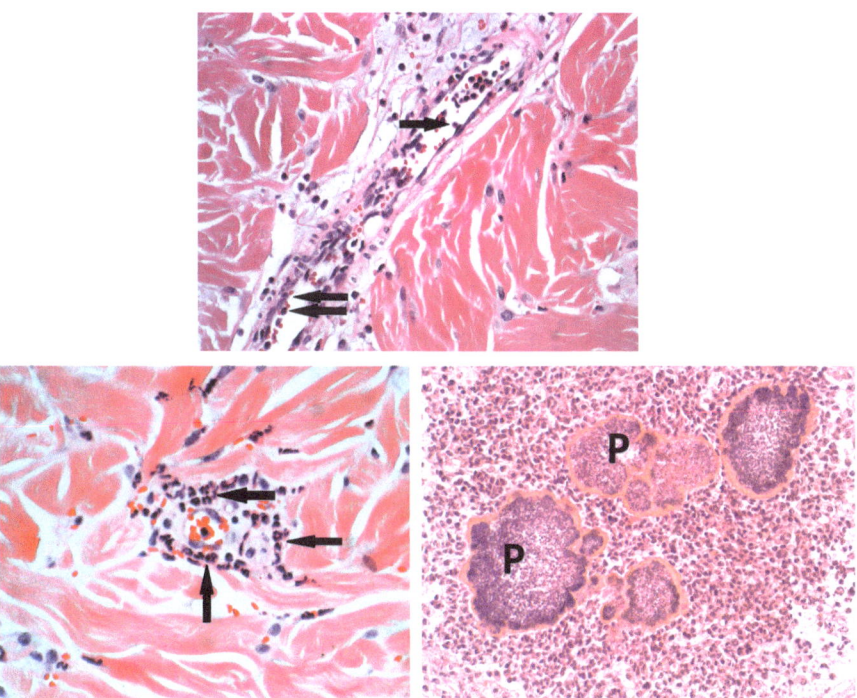

Figure 2.4 **Early inflammatory responses.** Distended capillaries in the skin (top panel), with leukocytes starting to marginate and migrate (arrows). This picture was taken about 6 h after the initial insult. In the bottom panel we move from the earliest inflammatory response on the left, with a few perivascular neutrophils (arrows), to the formation of pus in a well-established lesion (P) on the right.

2.2.2   Cellular Component

The cellular component can be divided into two phases. First the "wrecking crew" arrives; then the "clean up" crew takes over. The first phase is dominated by the neutrophils. Changes in the receptors on the inner surface of the capillary cause these cells to migrate and stick to the capillary endothelium. The first effect is described as margination, the second as pavementation (an awful word!). The polymorphs roll along over the endothelial surface, stick to it and then, very importantly, move through the endothelium. The beginning stages of this movement can also be seen in Figure 2.4. It seems likely that they move through the endothelial intercellular junctions; these become looser in an area where an inflammatory response is being mounted. Once outside the capillaries, the neutrophils go into action. Their main function is to attack bacteria: they are the shock troops of the response (it is difficult not to think in military terms when describing the acute inflammatory response). Whilst travelling in the blood, the neutrophils are well-behaved, rather quiet cells; as soon as they arrive in an area of infection or damage, they change their behaviour and become aggressive and attack bacteria. More chemical mediators are released by these activated cells, more cells are activated and more neutrophils are summoned. Many neutrophils die and, in breaking down, signal for further reinforcements. This is a positive feedback process: rather unusual in terms of control mechanisms in the body, which tend to involve negative feedback loops. In fact the response to injury is a cascade response, rather like the clotting cascade, and once begun is self-sustaining and self-reinforcing. It will be immediately obvious that such a response is a dangerous thing: it must be switched off before it runs out of control and starts doing more harm than good.

The neutrophil response to injury is very obvious in histological sections taken from inflamed tissue. The densely crowded nuclei of the neutrophils are clearly seen. If large number of neutrophils die in action, so to speak, their debris mixed with oedema fluid and protein produces pus. Both of these aspects are shown in Figure 2.4. This is characteristic of abscesses and less dramatic lesions, such as boils and pimples.

The second phase is dominated by the macrophage. These are the "clean up" cells and ingest not only bacteria but also debris left by the neutrophil attack; they can be seen in the top panel of Figure 2.5. Red cells that have escaped from damaged capillaries, dead neutrophils and damaged tissue cells are all ingested by the macrophages.

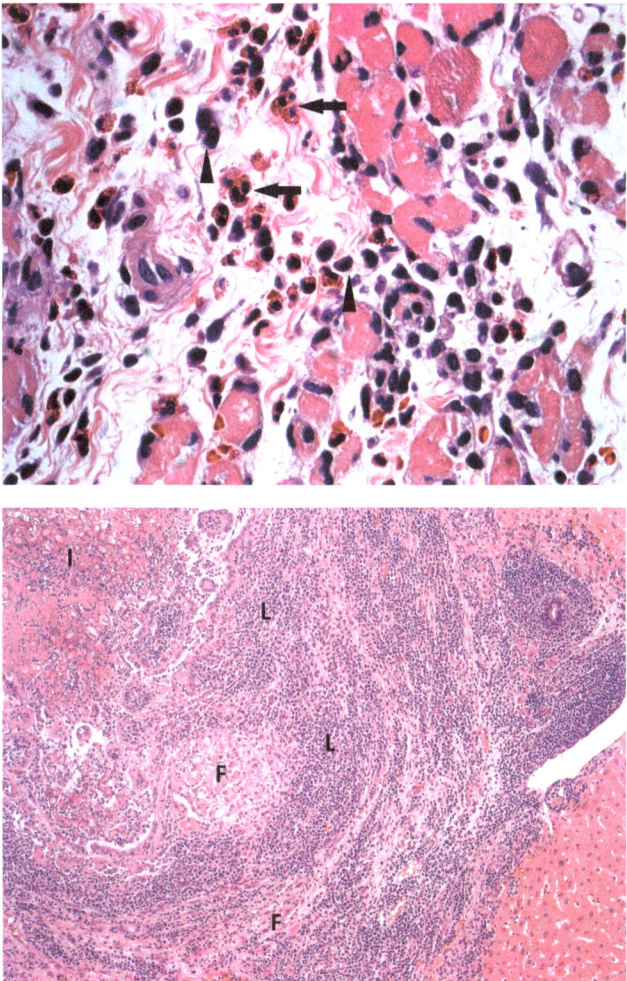

Figure 2.5 **Macrophages and chronic inflammation.** Macrophages at work: doing what they do best (top panel). This is skeletal muscle necrosis after injection of an adjuvant. The macrophages are full of bright eosinophilic particles (arrows) that are probably myoglobin from the muscle-fibre breakdown. There are also a number of mature plasma cells in the field (arrowheads). These are about the same size as the macrophages, with a purplish cytoplasm and small notches in the nuclei. Chronic inflammation in the liver (bottom panel). This response is the result of a parasitic Eimeria infection in the liver of a rabbit. The infectious agents can be seen mostly in the upper left quadrant (I). The response is predominantly lymphocytic (L), with a few macrophages and some early fibrosis (F).

Macrophages are large cells, much larger than the neutrophils (originally called microphages because of this) and are derived from the monocytes of the blood. They occur, again rather quietly, in connective tissue as histiocytes. As the cleaning up operation continues, repair begins. Restoration of the normal architecture of the tissue follows, but in many cases this is delayed or not fully achieved. One might describe the third phase of the cellular response to injury as the response of the fibroblast. Fibrosis, perhaps temporary, perhaps permanent, is often seen after tissue damage. We shall consider this further when we deal with repair processes.

2.3   CHRONIC INFLAMMATION

Cell damage leads to an acute inflammatory response. This is a defensive response and might be thought of as having evolved to deal with infectious organisms rather than toxic chemicals. The sequence of cellular destruction, the calling in of neutrophils to deal with invading organisms, the destruction of the invaders and then resolution of the lesion by healing seems, at least in teleological terms, a very sensible or appropriate response. If the invading agents are viral rather than bacterial, the neutrophil response will not be effective and a second wave of defence based on lymphocytes is necessary. Lymphocytes produce antibodies and attack and destroy cells that have been infected with viruses. If this line of defence is effective and the invading organisms are destroyed then the way is clear for the macrophages to clear the area of debris and for regeneration of damaged cells to occur. This process of repair is known as healing. Healing will be discussed later.

The rather clear cut sequence of events described above is sometimes disturbed. What if the invading organisms cannot be destroyed? Unsurprisingly the tissue response passes from one of rapid resolution to a chronic state. If military analogies are allowed, the mobile battle settles down to trench warfare. Damage continues to be done, a certain amount of repair occurs, reinforcements are called up, further damage is done and so the process grinds on and on with the tissue not being effectively restored to normal and the invader (perhaps an infectious agent, perhaps some inorganic material) not being effectively repulsed or destroyed.

Once the phase of acute inflammation has passed, the key cells involved in the chronic inflammatory response are lymphocytes: the small, round cells with dark-staining nuclei and hardly any cytoplasm that are characteristic of chronic inflammation; they can be seen in the bottom panel of Figure 2.5. Some chemicals produce damage to the liver, which leads to this sort of "viral-like response". This is characterised by lymphocytes accumulating in the connective tissue of the portal tracts. We might ask a few questions: what is the function of the lymphocytes that appear in the connective tissue of the portal tracts in this type of toxicological injury? Are they doing anything that might be described as "useful" in the sense of removing the chemical that is causing or has caused the damage, or are they responding to some signal from damaged hepatocytes that is similar to the signal sent by these cells in time of viral attack? The presence of plasma cells will be noted; these are the producers of antibodies *par excellence*; what are they doing with regard to injury caused by a chemical? None of these questions has been answered.

Chronic inflammation is produced by material that cannot be destroyed by the defending cells. Asbestos fibres and particles of silica provide good examples. The main attack on silica particles in the lung is led by the macrophages. These ingest the particles but are themselves destroyed by the particles. The process fails. Macrophages die and particles are released. More macrophages attack the particles. As damage occurs, inflammatory mediators are released and more and more defensive cells are called to the point of attack. At the same time, fibroblasts are stimulated and collagen is laid down. This, the laying down of collagen, can be seen as a further line of defence. Walling off the area of damage within a collagenous matrix occurs and fibrotic nodules are formed. All would be well if the tissue rested after having isolated the particles: perhaps they might be destroyed at leisure. Sadly, this is not always the case and the fibrotic response becomes progressive. It is perhaps not too much to say that acute exposure to very high concentrations of silica particles triggers a response in the lung that will eventually kill the patient. He dies of pulmonary fibrosis. The response of the lung to particles is an area of great interest: some particles, for example, coal dust particles, seem to be walled off but the fibrosis is not, in general, progressive; others, like silica, give rise to progressive nodular fibrosis; asbestos fibres cause a rather different sort of fibrotic response. The cells that mount these responses are, in all cases, similar; why different insults lead to subtly different pathological responses remains an unsolved puzzle. Presumably, the signals given off by the tissues attacked by these various particles must be slightly different. An example of granuloma formation in response to prolonged chronic inflammation in the lung is shown in Figure 2.6.

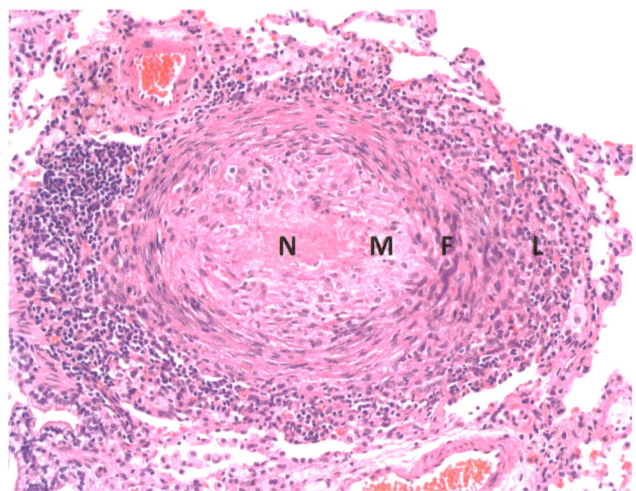

Figure 2.6   Granuloma formation in the lung. In some instances very long-standing
chronic inflammation fails to resolve and results in the formation of
granulomas. In the middle of the lesion is an area of necrosis (N)
surrounded by macrophages (M), with a thick layer of fibrous tissue (F)
and a few attendant lymphocytes (L).

The chronic inflammatory response might be described as frustrated healing. In teleological terms the tissue is "trying to heal" but
damage is still occurring. A combination of fibrosis, the extent of
which depends on the tissue involved and the injurious agent, and a
small, round cell infiltrate is standard. Macrophages involved in a
chronic inflammatory response often form giant cells. These, as the
name suggests, are unusually large cells and are characterised by
many nuclei: up to a hundred or so nuclei may occur in one giant cell.
Rather interestingly, there are several different types of giant cell: the
Langhans giant cell seen in the chronic foci of tuberculosis, the
Touton giant cell seen in the fatty xanthomata on the eyelids of some
elderly people and the foreign body giant cell seen clustering around,
for example, bits of insoluble suture material or particles in the lung.
These cells differ in the way in which their nuclei are distributed
throughout their cytoplasm. How very odd! What does it mean?
Nobody knows. Furthermore, nobody seems to know the function of
the giant cells: have they any function or are they "pathological" and
merely a sign of chronic inflammation? How do they form? They may
form *via* coalescence of normal macrophages or by replication of
nuclei within a macrophage, with expansion of the cytoplasm but no
cell division but, again, nobody seems to know.

2.4 HEALING OR REPAIR OF DAMAGED TISSUES

Acute inflammation is the inevitable consequence of injury. Even if the injury involves an incision made under aseptic conditions, some cells will be damaged and, as they undergo necrosis, they will call in, so to speak, the acute inflammatory response. The incision will also have divided blood vessels and the extravasated blood will have clotted in the wound. In an uninfected incision the acute inflammatory response is brief and healing begins rapidly. Most of the classical work on healing processes was done with respect to wounds: lacerations and fractures of bones. A great deal, for example, is known about the healing process of bones, but this is not of great importance to toxicologists. The process of healing is, however, similar in all tissues. The essential steps are the removal of debris, including extravasated blood and the fibrin that has been deposited in the damaged area, construction of a connective tissue scaffolding, simultaneous invasion of the area of damage by blood vessels and the development of a firm connective tissue linkage across the damaged area. In skin the firm linkage between the undamaged tissues, the bridge across the damaged area, forms a scar. An uninfected incision heals rapidly and the scar is narrow. The collagen fibres that form the bridge, and thus the scar, gain strength as the healing process matures: the rate of gain of strength is, in part, dependant on the stress placed on the tissue (very important for the normal healing of fractures) and, also importantly, on factors such as vitamin C. Wounds in patients suffering from a lack of vitamin C heal very slowly.

Infected wounds heal more slowly. Surmounting the bacterial invasion can be a slow process made slower by, for example, a poor blood supply to the affected area. The infected tissue represents a battle ground where bacteria are multiplying and being destroyed, waves of inflammatory cells are being deployed, with many dying and forming pus if the bacteria are of certain types, macrophages removing damaged tissue, blood vessels invading the damaged area and fibroblasts forming collagen. This messy mixture is described as granulation tissue: if the pus and debris is cleaned away from the surface of an infected wound (for example by use of a swab), the surface will be seen to be granulated with lots of tiny pink granules and points of red produced by bleeding capillaries. Healing eventually occurs but the fine scar produced when an uninfected wound heals quickly is replaced by a larger scar that takes longer to gain strength and may breakdown again.

Tissues characterised by a rapid turnover of cells heal well. The lining of the mouth for example heals rapidly: the surface epithelial cells are rapidly replaced and an abrasion of the surface is rapidly closed. Healing or repair in such tissues is a process of restoration of the pre-injury state and, when healing is complete, it will be difficult to see that an injury has occurred. Tissues like liver, in which the cells (hepatocytes) are not normally turning over as rapidly as those of the gut or skin, are also able to heal to a normal state by stepping up the rate of cell division. The central nervous system, on the other hand, undergoes a form of repair in which the supporting cells (glia) form a scar, but the neurons lack the capacity to divide and rebuilt the normal structure of the tissue.

In a tissue like liver, healing requires three processes to occur in a coordinated way. Scaffolds of connective tissue and blood vessels must be laid down at the same time as the parenchymal cells (the hepatocytes) are dividing and taking their places within the scaffold and also at the same time that necrotic material is being cleared away by macrophages. Consider a small volume of liver that has been damaged by, say, a chemical. Cells die within the damaged zone and inflammation follows. Assuming that the chemical is removed (how this occurs will be discussed later but, for the moment, let us say that it has been broken down to less harmful materials or has been excreted by the kidney), the repair process can get underway. The necrotic tissue will be removed by macrophages, and fibroblasts will begin to lay down connective tissue. Reticulin fibres are the first to be formed, followed by collagen. There are many subtypes of collagen; the type involved in repair processes is Type III collagen. If the process is followed by the use of special stains then the fibrin that will be seen early on in the process will be seen to be replaced by collagen. The parenchymal cells will be seen to be dividing (many mitoses will be visible) and organising themselves into the normal liver pattern. Restoration of normality is underway. But this process can fail. Overproduction of connective tissue can lead to islands of regenerating liver being surrounded by dense walls, so to speak, of collagen. What has happened is a failure of coordination and control. This is especially likely if the damage is repeated by repetitive exposure to the injurious chemical. The cirrhotic liver produced by the overconsumption of alcohol represents just such a picture. Nodules of regenerating liver parenchyma, surrounded by fibrous tissue, appear. The exact distribution of the damage and of the repair process depends on the chemical producing the damage. It is hardly surprising that a tissue that is damaged over and over again fails, in the end, to regenerate its normal structure. The phrase "frustrated regeneration" neatly expresses the problem.

2.5   HYPERTROPHY, HYPERPLASIA AND METAPLASIA

2.5.1   Hypertrophy

Every schoolboy and body-builder knows that skeletal muscles can be enlarged by programmes of exercise often involving isometric exercise. Rather less obviously, the left ventricle of the heart enlarges if the aortic valve is incompetent and it has to pump more blood than usual. Similarly, the right ventricle of the heart enlarges when the pressure in the pulmonary circulation rises. The uterus enlarges during pregnancy. A final example: the smooth muscle of the bladder enlarges in old men with urethral obstruction caused by an enlarged prostate gland. In each of these cases, if histological sections of the organs were examined, it would be seen that muscle cells had increased in size. Looking more closely we would see that the number of cells present (comparison with a control section would of course be needed) had not changed and the number of mitoses was as usual: very few. The changes described here are those of hypertrophy. The word hypertrophy is used in a special sense by histopathologists. It is used to indicate that cells have become enlarged; it does not apply only to muscular tissue and an example of hypertrophic liver is shown in Figure 2.7.

The response, in muscle, is to mechanical work and presumably increases the number of contractile elements available in each cell. As the cells enlarge, the centre of the cells lie further from the cell surface and from the capillaries, and thus the supply of oxygen is diminished. Ischaemic pain from the hypertrophied left ventricle is common-place. What is a little confusing is that the word hypertrophy is also used, but not by histopathologists, to refer to enlargement of an organ wherein the cells may be of normal size but present in much greater numbers than is usual. This change is described as hyperplasia.

2.5.2   Hyperplasia

Hyperplasia occurs only in tissues where the cells are capable of replication. For example, the glandular tissue of the female breast enlarges during pregnancy and the breasts themselves become larger. This is a physiological effect of hormonal stimulation. It would be right to refer to the glandular tissue as hyperplastic; one would not refer to it as hypertrophic because the cells remain of normal size. One might say that the breasts had undergone a hypertrophic change,

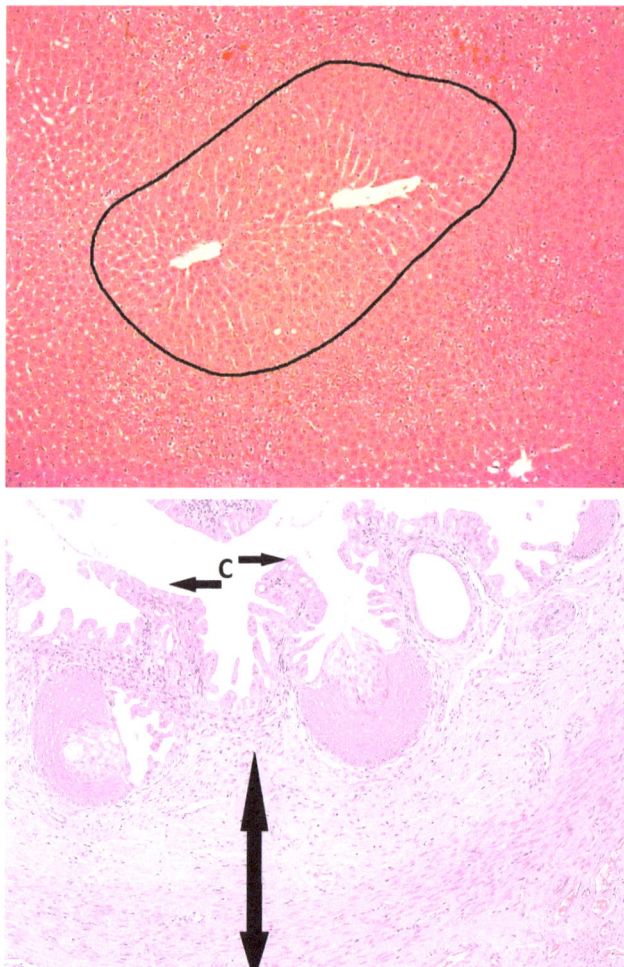

Figure 2.7 **Hypertrophy and metaplasia.** The top panel shows an example of liver hypertrophy. The affected areas are within the outline. In the bottom panel we have squamous metaplasia in the uterus. Although the columnar epithelium (C with arrows) is not quite normal, the areas of squamous metaplasia with their stacked layers of cells can be seen clearly (double-ended arrow). This was not as a result of irritation, but was the result of a hormonal imbalance.

but this would be to use the term with reference to the whole organ rather than to its cells. Other glands also enlarge under hormonal stimulation.

Another example is provided by the kidney. If one of the two kidneys is removed, the remaining kidney enlarges. Common sense tells us that it has enlarged to allow it to cope with the work of two kidneys.

This does not take us very far: what are the signals that trigger the hyperplasia (for this is not cellular hypertrophy)? Is it the increased blood flow, the increased volume of filtrate passing across the glomeruli, the increased amount of, for example, glucose being re-absorbed by the tubules or what? Where are the transducers that receive the signals and convert them into instructions to the cell to divide? And what are those instructions? Payling Wright, in his excellent work *Introduction to Pathology* published in 1950 but still readable and instructive, discussed the comparative effects of extra protein and urea on renal hypertrophy.[†]

The key finding was that increasing the nitrogen load causes an increase in the size of the kidneys. It was also noted that the effect seemed to be due to more than just the nitrogen load: when urea was used to increase the nitrogen load the effect on the kidneys was not as marked as when the increased load was induced by an increase of protein. Payling Wright suggested that the extra work involved in deamination of amino acids might be an important stimulus. But precisely how this stimulus is received remains obscure. Note that Payling Wright used the term hypertrophy: we would agree that the kidneys were hypertrophic but would point out that the cellular change was one of hyperplasia.

2.5.2.1 Stability of Cell Populations. In thinking about hyperplasia and hypertrophy it is helpful to divide cells into three groups. This division is perhaps not as absolute as once was thought but it is still helpful.

1. **Labile cells:** Those undergoing replication all the time; for example, the cells of the epidermis and the lining epithelium of the gut.
2. **Stable cells:** These turn over much more slowly than the previous group: a long rest period between cell division and replication of DNA occurs. In some tissues it is difficult to see any evidence that the cells are dividing: mitoses are few and far between. But these cells can divide if necessary; the cells of the liver provide an excellent example. Mitoses can be found in histological sections of the normal liver but an extended search is likely to be necessary before one will be found. Damage to the liver, for example by alcohol, provokes bursts of cell division and regeneration occurs.
3. **Permanent cells:** In these cells, division does not occur, at least not after birth. Neurons and muscle cells (all types) are the

---

[†]Wright G Payling, *An Introduction to Pathology*, Longmans, Green and Co., London, 1950.

classic examples. We have seen that muscle cells respond to mechanical stress by hypertrophy; mental effort does not produce hypertrophy of neurons (fortunately, given the presence of the skull). Of course mental effort does improve mental performance, but this is achieved by other mechanisms.

2.5.3   Metaplasia

The conducting airways of the lung are lined with ciliated epithelium. The epithelium is actually rather complex and is referred to as a pseudostratified columnar ciliated epithelium. Recent work has suggested that this long-established description might not be entirely accurate and that the epithelium is actually stratified with not all the cells reaching down to the basal lamina. Chronic irritation and damage to the epithelium, produced for example by smoking, leads to distinct changes in its structure. Areas of squamous epithelium appear: this change is described as squamous metaplasia, an example from uterine tissue is shown in Figure 2.7. The epithelium has changed from one well-differentiated form to another. The ciliated epithelium can also change into a columnar epithelium: one without cilia but with the capacity to secrete mucus. In addition, an increase in the number of goblet cells can occur. Both changes are seen in chronic bronchitis.

Squamous metaplasia might appear to be "useful": the normal epithelium has been replaced, at least in part, by one better able to cope with "wear and tear". This is in fact not the case and metaplasia of any sort is a rather sinister change that may presage malignant changes of the cells.

2.6   TUMOURS: BENIGN AND MALIGNANT

Only a very brief introduction to the histopathology of tumours is provided here; no discussion of the mechanisms of mutagenesis and carcinogenesis is provided as these are dealt with in detail in textbooks of toxicology.

What is a tumour? The word itself means a swelling, but that does not take us very far: there are many causes of swellings. Is a tumour a cancer? Yes, in many cases it is, but the common wart is a tumour of a kind—a benign tumour that does little harm and few would describe a wart as a cancer. No completely satisfactory definition of a tumour has been produced, though most people and all toxicologists know what is meant by the term. In general terms a tumour is a mass of cells that has escaped from the control mechanisms that normally limit the growth of tissue. Thus, a wart grows as a result of a viral infection and forms a swelling. The viral infection has caused the epidermal cells to escape from the normal mechanisms of control of cell division. The cells of a breast cancer have also clearly escaped from the normal mechanisms of control of cell division, but have also acquired characteristics very different from those of a wart. For example, they have acquired the capacity to invade normal tissues, to spread though the vascular and lymphatic systems and to establish secondary tumours (metastases) in other tissues, such as the liver and the brain. They also multiply very much faster than those of a wart. Breast cancer is a lethal disease and the tumour is rightly described as malignant. The wart, on the other hand, causes few problems and is described as a benign tumour. The term neoplasia, meaning new growth, is often applied to the process of tumour development. Neoplasia implies more than hyperplasia (the abnormal replication of normal cells) and metaplasia (the conversion of one type of cell into another, often with an element of hyperplasia). In neoplasia a new type of cell is produced. New, that is, in the sense of not being under normal control; it may in fact look not particularly abnormal, although many look distinctly abnormal. One more difference from hyperplasia is important: hyperplasia stops and the tissue returns more or less to normal when the stimulus to hyperplasia is removed; neoplasia persists. For example, a skin tumour induced by tar does not regress when application of the tar is stopped; on the contrary, it continues to grow.

Tumours may be classified in a variety of ways. We have considered one: the division into benign and malignant. Table 2.1 is long and introduces a large number of new words. It is provided for reference but is invaluable for understanding the dual classification of tumours: by tissue of origin and by behaviour.

Already we are loaded with new words! Note that papilloma implies a benign epithelial tumour; while adenoma means a benign tumour of glandular epithelium. Malignant tumours arising from epithelia are described as carcinomas, from glands they are adenocarcinoma. Adenomas of the lung are very common in some strains of mice; adenocarcinomas are induced by many carcinogenic chemicals.

Note that sarcoma is used for malignant tumours arising from tissues other than epithelium.

Of course, tumours could also be classified by their precise tissue of origin, for example, an adenocarcinoma of the lung or of the pancreas or of the parotid gland. In addition to all this there are sub-classifications and subtypes: the list is endless. Learning to recognise

Table 2.1   Characteristics and behaviour of tumours.

Tissue of origin	Behaviour		
	Benign	Intermediate	Malignant
Epithelium covering epithelia			
Squamous	Squamous cell papilloma		Squamous cell carcinoma
Transitional	Transitional cell papilloma		Transitional cell carcinoma
Columnar	Columnar cell papilloma		Adenocarcinoma
Compact secreting epithelium	Adenoma. If cystic: cystadenoma or papillary cystade-noma if containing fronds of tissue		Adenocarcinoma; if cystic: cystadenocarcinoma
Other epithelial cells		Basal cell carcinoma, salivary and mu-cous gland tumours, carcinoid tumours. These are difficult to classify in terms of behaviour	

Tissue of origin	Behaviour		
Connective tissue	Benign	Intermediate	Malignant
Fibrous tissue	Fibroma		Fibrosarcoma
Nerve sheath	Neurofibroma		Neurofibrosarcoma
Fat	Lipoma		Liposarcoma
Smooth muscle	Leiomyoma		Leiomyosarcoma
Striated muscle	Rhabydomyoma		Rhabydomyosarcoma
Synovium	Synovioma		Malignant synovioma
Cartilage	Chondroma		Chondrosarcoma
Bone			
Osteoblast	Osteoma		Osteosarcoma
Osteoclast	Osteoclastoma		Malignant osteoclastoma
Mesothelium	Benign mesothelioma		Malignant mesothelioma
Blood vessels and lymphatics	Benign hae-mangioma, lymphangioma		Angiosarcoma
Meninges	Meningioma		Malignant meningioma

Table 2.1 (*Continued*).

Tissue of origin	Behaviour		
	Benign	Intermediate	Malignant
Specialised connective tissue			
Neuroglia	Astrocytoma, oligodendroglioma, ependymoma: all are forms of glioma	The classification with regard to behaviour is difficult here: some gliomas grow rapidly, and spread outside the brain is variable. Note that these often malignant tumours are described as gliomas and not as "gliosarcomas"	
Chromaffin tissue	Carotid body tumour		Malignant tumour of the carotid body
Haematopoietic and lymphoid tissue	Benign lymphoma		Lymphosarcoma, reticulum cell sarcoma, Hodgkin's disease, follicular lymphoma, leukaemias, Polycythaemia rubra vera
Melanocytes			Malignant melanoma
Foetal trophoblast	Hydatidiform mole		choriocarcinoma
Embryonic tissue			
Totipotent cell	Benign teratoma		Malignant teratoma
Multipotent cell			
Kidney			Nephroblastoma
Liver			Hepatoblastoma
Unipotent cell			Retinoblastoma
Retina (further examples not included here)			
Embryonic vestiges			Chordoma
Notochord			
Hamartoma: see below			

a range of tumours can only be accomplished by experience. Looking at the pictures in atlases helps.

A little more detail concerning benign and malignant tumours is given in the following sections.

2.6.1   Benign Tumours

Benign tumours do not metastasize. This is a golden rule. In addition, the cells of a benign tumour closely resemble those of the parent tissue and thus they are defined as "well differentiated", as shown in the top panel of Figure 2.8. Benign tumours can grow to large sizes but they do not invade other tissues. Any clinical effects that they have are due to pressure, obstruction, *etc*. Death from a benign tumour is unusual. A good example of a benign tumour that can grow to large sizes is the uterine leiomyoma. As the tumour grows, the muscle cells may be replaced by connective tissue cells and much collagen is laid down. The tumour is then described as a fibroleiomyoma (these terms

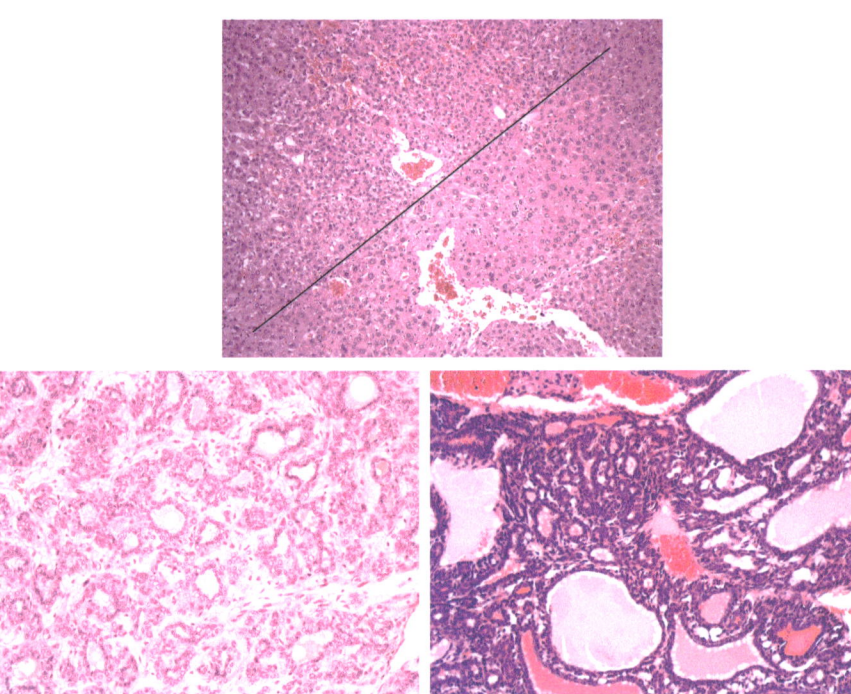

Figure 2.8   Benign and malignant tumours. Benign liver tumour (top panel). The tumour cells are those in the top left half of the panel above the line. If you look closely, you can see that the nuclei are smaller and more basophilic, but not obviously a great deal different from the normal hepatocytes in the bottom half. Benign tumour compared to malignant (bottom panels). On the left is a benign mammary tumour from a rat. In this case it looks reasonably different from normal mammary tissue, but is still recognisable. To the right is a malignant tumour of the same origin. Note the intense basophilia of the malignant tumour—a reflection of the very high DNA content—and the piling up of the epithelium.

are long!) or fibroma for short. As the benign tumour grows, the normal tissue around it is compressed and may form what looks like a capsule. The tumour does not spread outside the capsule. No capsule is formed by papillomas (or papillomata for classicists).

Although the cells look much like the parent cells, we should be unsurprised to see more mitoses than usual though not as many as we might have expected. Benign tumours can change their behaviour and become malignant; an example of this transition can be seen in the bottom panels of Figure 2.8. The splendidly named dermatofibrosarcoma protuberans, meaning a skin tumour that began as a fibroma, became malignant and protrudes from the surface, is a good example. I recall caring for a woman with this awful tumour. On two occasions over a period of about twenty years, a fibroma had been removed from the patient. On each occasion, the first presentation and the recurrence, the pathology report described the tumour as benign. When seen by the author the patient was suffering from multiple and widespread metastases; she died rapidly and the post-mortem showed large tumours throughout most of the tissues and organs of the body. Metastases to skeletal muscle are rare but this patient had many. Sections revealed a malignant fibrosarcoma. Such tumours are fortunately rare.

The basal cell carcinoma of the skin is a tumour that does not metastasize, but which erodes local tissues. It is known as a rodent ulcer because of this propensity to eat into normal tissue and the older textbooks of pathology show appalling pictures of patients, where half the face had been destroyed and yet in whom no metastases had occurred. Such advanced cases are not seen in the UK today as early diagnosis is standard and surgical treatment is curative.

In addition to looking like the cells of the parent tissue, benign tumours may function like the parent tissue. Thus, tumours of endocrine glands may produce hormones in large quantities. This may have secondary effects. For example, a tumour of the adrenal medulla, a phaeochromocytoma, may produce enough adrenalin and/or noradrenalin to produce paroxysms of hypertension.

2.6.2 Malignant Tumours

Malignant tumours are potentially lethal and often metastasize. To the general public, "cancer" means a malignant tumour. The spread of malignant tumours into the surrounding tissues may look, at post-mortem examination, like a crab: hence the word "cancer". As malignant tumours invade tissue, the invaded tissue is destroyed and

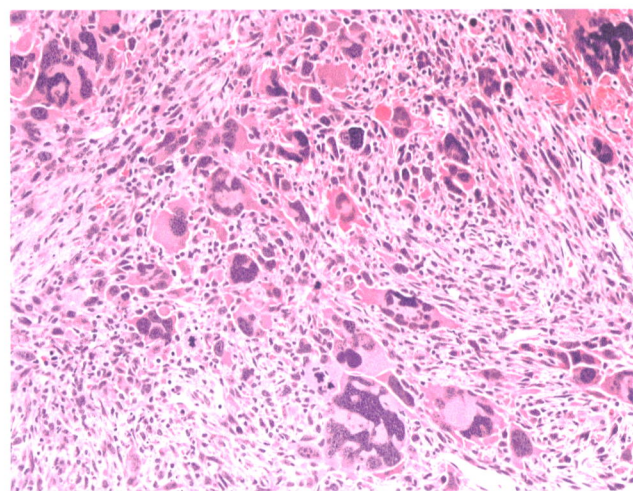

Figure 2.9   Malignancy. A malignant subcutaneous tumour from a mouse. It is
practically impossible to tell the tissue of origin. Some of the nuclei
and mitotic figures are as bizarre as anything you are ever likely to see.

no capsule of attenuated normal tissue is usually found. The cells of
malignant tumours differ much more from those of the parent tissues
than is the case with benign tumours. The degree of differentiation
also varies: in some malignant tumours the cells appear remarkably
undifferentiated and are described as anaplastic. Classifications
based on the degree of differentiation have been produced. The cells
of malignant tumours tend to vary in size (pleomorphism) and abnormal mitoses are often seen. A jumbling of large cells and small
cells, cells with normal nuclei and cells with very abnormal nuclei is a
common picture. An example of the variable cell types and nuclei can
be seen in Figure 2.9. In addition, the cells of a malignant tumour lose
the tendency of normal tissue cells to stay together: cells leave the
main mass of the tumour and spread into blood vessels and lymphatic vessels. Thus, clumps of tumour cells may be found in veins, and
secondary tumours (metastases) are often found in lymph nodes.
Tumour cells also spread along fascial planes. Classifications based
on the extent of spread of a tumour are often used in clinical practice.

Table 2.2 sums up the differences between benign and malignant
tumours.

2.7   SPECIAL TERMS

Teratoma: A teratoma is a tumour arising from embryonic tissue and
contains a variety of tissues. They are often found in the ovary or

Table 2.2 Characteristics of benign and malignant tumours.

Benign tumours	Malignant tumours
1. Growth	
Slow growing	Rapidly growing
Expansive type of growth	Invasive type of growth
Progression is erratic with a tendency to cease	The progress is usually relentless until death occurs.
2. Metastases never occur	Metastases frequent
3. Size	
Usually of small size but occasionally enormous	Usually of large size
4. Histological structure	
Well differentiated	Less well differentiated and sometimes completely anaplastic
Well-formed stroma with little tendency to haemorrhage and necrosis	Stroma often poorly formed. Haemorrhage and necrosis common.
Cells regular, few mitoses	Cells often plaeomorphic, mitoses often numerous.
5. Cause of death	
Usually not fatal. If death occurs, it is due to mechanical effects. Endocrine adenomata may have hormonal effects	Almost invariably fatal if untreated. Cause of death: a combination of mechanical and destructive effects, together with blood loss, secondary infection, starvation *etc.*

testis, and less often in the mediastinum or inside the skull. Such tumours often contain a jumble of tissues: bone, cartilage, teeth (!), hair and cysts lined with gut or respiratory epithelium are often seen.

Hamartoma: A Hamartoma is a tumour-like malformation containing the tissues of the organ or the part in which it arises but arranged in a jumbled way. Thus, an isolated mass of cartilage containing clefts lined by respiratory epithelium may be found in the lung. Many more examples are known, including the pigmented naevi of the skin.

2.8 CONCLUDING REMARKS

If this information seems overwhelming, remember this section was meant to be an overview to histopathology in general and many aspects may not apply to your particular study. Feel free to filter this information as we move forward into the more generalised areas of histology.

CHAPTER 3

The Light Microscope

The light microscope is the essential tool of the histologist. All histologists need to know how to get the best from their microscope;
this is not particularly difficult to achieve, especially if modern instruments are available. Despite this, many histologists show little
interest in microscopy and put up with poor performance from
their microscopes. This poor performance can be corrected if the
principles on which a microscope works are understood. In this
chapter we start at the very beginning and develop the theory of the
light microscope in easy stages. At every point, emphasis has been
placed on practicality: how to use the microscope in addition to
understanding how it works. Electron microscopy has not been included. This is because it involves techniques that are often beyond
the reach of the beginner; whilst most biologists have access to
their own light microscope, few, even now, have access to an electron microscope. Immunohistochemistry is considered in a later
chapter of this book and a short description of the light microscope
as adapted to take advantage of fluorescence as a means of identifying components of tissues and cells is included here.

3.1 INTRODUCTION

The light microscope is perhaps the instrument most widely used in
biological research; it is also, rather surprisingly, one of the least well
understood by its users. Perhaps even more surprising is the fact that

the majority of histologists and pathologists seem not very interested in microscopy.

All biologists are familiar, to a greater or lesser extent, with the light microscope (hereafter referred to as the microscope): they may have been given one as a child; they will have used one in school and at university; and they may have used one in their research work. Modern microscopes are easy to use and, unless the instrument is very badly adjusted or has become damaged in some way, the user will see something of the specimen being examined. An example of a high-quality standard microscope is shown in Figure 3.1.

The quality of the image produced may, however, be rather un-satisfactory and the user may be disappointed. As soon as the image is photographed, its defects may become more obvious and the user may wonder why his or her photographs (photo-micrographs) are not as good as those he or she sees in textbooks or at scientific meetings. The answer is likely to be that the microscope is not being used as well as it could be. This chapter explains how a microscope works and how it should be used to produce the best performance of which it is capable. The explanations relate to the use of the microscope in what is described as "bright field" mode. Once this is mastered, the techniques of dark field, phase contrast microscopy and fluorescence microscopy follow on fairly easily; an example of their usefulness is shown in the bottom panels of Figure 3.1.

All biologists know that the microscope works by forming an en-larged image of the specimen being examined. The specimen, for our purposes a histological section mounted on a glass slide beneath some transparent medium and a coverslip, is illuminated by a light source. Rays of light leaving the object enter the objective lens of the microscope and are brought to a focus within the tube of the in-strument. The image that is produced is real (real in the sense that if a screen could be introduced at this point the image would be thrown upon it), enlarged and inverted. This primary image forms the object for a second lens, the eyepiece or ocular, and a secondary image is produced. This secondary image is a virtual image; it is enlarged and upright. The secondary, virtual image has no reality in the sense that it cannot be caused to appear on a screen, but light that seems to come from this image enters the eye and forms an image on the retina. It is important to remember that this final image is produced by the refracting system of the eye (the curved cornea and the lens). The secondary image lies at more than twice the focal length of the eye (meaning the refracting system of the eye) from the eye. The image produced by the eye is thus inverted, in terms of the secondary image,

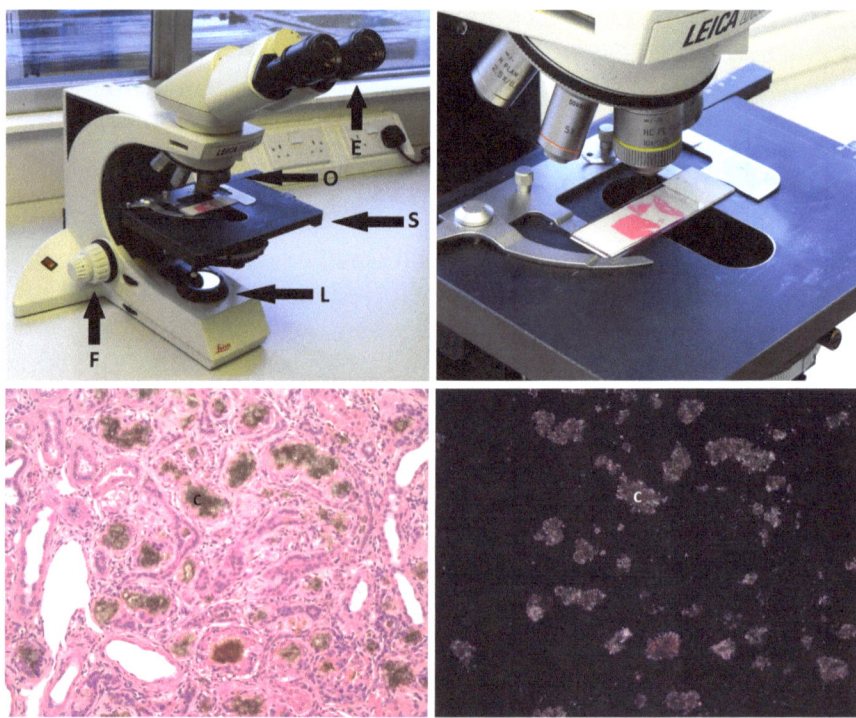

Figure 3.1 **High quality light microscope.** You don't have to go to this level, but a quality microscope with six objectives (O), ranging from 2.5× to 63× dry, gives you a good chance of identifying the majority of changes you may come across. Other standard parts of the microscope are the eyepieces (E), the stage (S), where the slide is held in place and illuminated for visualisation, and the focus controls (F), where the outer ring controls the coarse focus and the inner is the fine focus. The illumination on this microscope comes from a lamp and is reflected up into the sample *via* the lens (L) under the stage. The top right panel shows a closer view of the objectives. Everything you need to know is written on the objective lens. The bottom panels show that not all light is equal. Although these crystals (C) are clearly visible under normal white light (bottom left), polarized light reveals them in all their glory (bottom right).

and diminished in size. This final image is rather small: it has to be to fit, so to speak, on the retina, as indeed the image of St Paul's Cathedral fits onto the retina when we look at it from the Thames. It looks, however, very large in the sense that the brain interprets the image as being formed by light coming from a large object. The original object is thus seen as both magnified and inverted. There is room for confusion here! The final image on the retina is actually the same way up as the original object. But it will appear inverted because the brain inverts, so to speak, all images formed on the retina. To be clear:

- The object is upright;
- The primary image is inverted in terms of the object;
- The secondary image is upright in terms of the primary image, thus inverted in terms of the object;
- The final image is inverted in terms of the secondary image, thus upright in terms of the object;
- The brain inverts the final image, and thus "sees" the object as inverted.

3.2 RAY DIAGRAMS

Discussions of the paths taken by rays of light passing through the microscope are aided by ray diagrams.

When drawing a diagram of the paths taken by rays through a lens system two simple rules must be remembered:

1. Light travelling from any point on the object through the optical centre of the lens is not refracted, that is, it is not deflected from its original path.
2. Light travelling to the lens parallel to the axis of the system is refracted *via* the principal focus of the lens.

These rules allow the position and size of the image produced by any lens to be defined. They also allow the vertical orientation of the image to be defined. The lines representing these rays are known as the lines of construction. They represent only two of the rays coming from any one point on the object. Figure 3.2 shows the path taken by light passing through a compound microscope. Note the primary image and the secondary image. Note too that the working of a compound microscope can be best understood by considering the objective lens and then the eyepiece and then putting them together.

All this could be achieved without a microscope as such. The reader may recall the optical bench from classes in elementary physics. Here, two simple biconvex lenses are mounted on a horizontal set of rails: no microscope tube is needed. Let us assume that the object is luminous; for example, a candle flame. Light from the flame passes through the first lens (the objective lens) and then through the eyepiece lens. If the flame is examined *via* the eyepiece lens, an enlarged and inverted image of a part of the flame will be seen. The primary

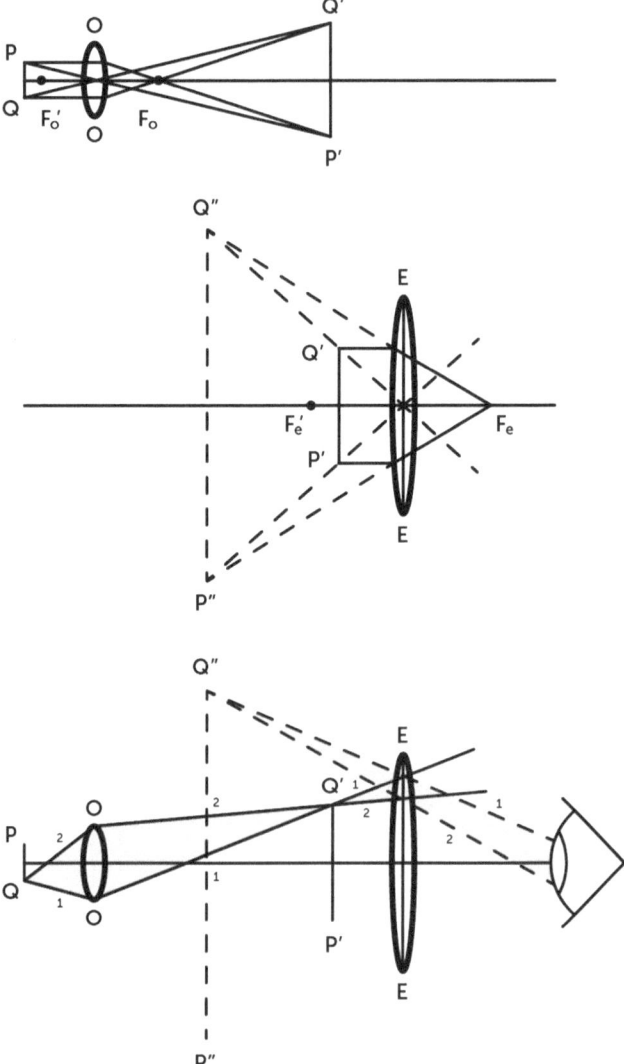

Figure 3.2 Diagrams illustrating how the image is built up in a compound microscope. OO: objective lens, EE: eyepiece lens, PQ: object, Q'P': primary image, Q"P": secondary image. Fo: focal point of objective lens, Fe: focal point of eyepiece lens.
Reproduced, with permission, from *Lecture Notes On Microscopy*, R Barer 1971.

image formed by the objective lens has acted as an object for the eyepiece lens. The eyepiece lens has formed a virtual image of the primary image and it is light that seems to come from this secondary image that is focused by the eye.

If a piece of cardboard is moved about between the two lenses, the image produced by the objective lens can be demonstrated. Of course, the image of the flame produced by the objective lens cannot be seen without the aid of the screen. The primary image is described as an "aerial image": this means that it is present in the sense that the light from the object is brought to a focus at the point where the image can be seen if a screen, the piece of card for example, is introduced. The object lies just outside the principal focus of the objective lens, and thus the real, enlarged and inverted primary image is formed. The primary image, formed by the objective lens, lies just inside the principal focus of the eyepiece lens, thus the virtual, enlarged and upright secondary image is formed: see Figure 3.2. This secondary image has no reality in the sense that it cannot be caused to appear on a screen. Light that seems to come from this secondary image requires a further "lens", the refracting system of the eye, to be brought to a focus on the retina. When the optical system of the eye is relaxed objects at infinity are focused on the retina; this means that the retina is at exactly the principal focus of the eye's refracting system. The refracting system of the eye is focusing parallel rays coming from the object at infinity onto the retina. To focus any object brought closer to the eye and from which rays of light will be diverging, the physiological process of accommodation is needed: the focal length of the refracting system of the eye is reduced and the image remains sharply focused on the retina. There is of course a limit to how much the focal length of the eye can be reduced: the near point is reached in young people at about four inches from the eye. Aging reduces the power of accommodation and the near point becomes further from the eye. People with normal eyes read well when print is placed at a distance of about 25 cm from the eye. Some accommodation is, of course, needed to focus objects at this distance. The lens system of the microscope is adjusted so that the secondary image is produced at about 25 cm from the eye and can thus be viewed without strain to the eyes.

The eyepiece can be used to form a real image of the object, which can be projected onto a screen. To do this, the eyepiece lens will have to be moved so that the primary image, the object of the eyepiece lens, lies just outside the primary focus of the lens. If the lenses are arranged in this way then a piece of cardboard, the screen, can be placed at about where the eye would be placed and an image of the flame will be seen on the screen. The system is now working as a projection microscope. This is how the microscope is set up for photography: the image is projected onto photographic film. When a modern microscope is used for photography, an additional lens is incorporated so that the object can be viewed as usual, whilst part of the light (the beam is split by a prism) is focused on the photographic film.

If the original object is not luminous, some source of light will be needed. If the candle flame is replaced by a histological slide and lit by a lamp placed beyond it, the section will act as a luminous object (light from the lamp passes through it and a series of images will produced, as described above). The image is produced by transmission of light through the object: transmission light microscopy. If the flame is replaced by something that is not transparent, for example a pin, then the image will reveal only a shadow of the pin: no detail of the surface of the pin will be revealed. Such detail could be provided by moving the light source so that the light is reflected from the surface of the pin onto the objective lens. This is described as incident light microscopy or as reflected light microscopy.

All the major image-forming components of the light microscope have now been described. What remains are refinements needed to produce a high-quality image. Each component of the microscope will be considered in turn.

3.3   THE OBJECTIVE LENS

The objective lens (the objective) is the most important component of the microscope. In the description given above and in the ray diagram, we considered only a simple biconvex lens. Such lenses were used in early microscopes but were found to be unsatisfactory because of aberration. What does this mean? All lenses fail to produce an image that is an exact replica of the object being viewed because of three things.

Light passing through the edges of the lens (those parts far away from its optical centre) are refracted by the lens to a greater extent than those passing nearer its centre, and are thus brought to a focus nearer to the lens than the rays that passed closer to the axis. This produces blurring of the image: a form of aberration described as spherical aberration. Microscope manufacturers correct objectives for spherical aberration by using combinations of lenses rather than just one lens. The calculations of just what adjustment is needed are complicated and involve "the lens computer". Such calculations make assumptions about the objects being examined. One factor that is critically important in histological work is the thickness of the coverslip (and mounting medium). We shall return to this point.

The second type of aberration is chromatic aberration. The refracting power of a lens depends on the wavelength of the light being refracted. It would be more accurate to say that the refractive index of the glass depends on the wavelength of the light. Blue light is more strongly refracted than red light. White light contains a mixture of wavelengths and light of these various wavelengths cannot be brought to exactly the same focus by a single lens. Correction is achieved by incorporating lenses made of different sorts of glass into the objective. This brings light of differing wavelengths to a single focus. Lens can be corrected for a variable number of wavelengths. The achromatic objective introduced in about 1820 was corrected for two wavelengths. The apochromatic objective, introduced by Abbe in 1886, allowed correction for three wavelengths and was a most important step forward. Needless to say, apochromatic lenses (apochromats) are much more expensive than "achromats" as a result of their greater complexity.

The third form of aberration suffered by simple lenses is due to the different distances that light rays must travel when passing through the edges as compared with the centre of the lens. This produces a curved image. Thus, though the object (the histological section in our case) may be flat, the image will be curved. This was a significant drawback of older, otherwise well-corrected, apochromatic objectives. On viewing the image it will be found that, when the centre of the image is in focus, the edges are not. This does not matter too much for ordinary observation: one looks at the centre of the field and moves the slide about bringing different areas of the specimen (in our case the histological section) into sharp focus. But for photography the curved field is a real problem: the centre of the photograph will be in focus, the remainder will not. A solution is provided by flat field objectives. These complicated lenses correct for field curvature by incorporating more lenses into the objective. The most expensive lenses available are described as plan-apochromatic and are corrected for spherical and chromatic aberration and for field curvature. The cost of a top quality plan-apochromat oil immersion lens might approach £2000. The author was allowed to buy one costing £1400 more than twenty years ago. Such lenses do not offer better resolution than standard apochromats. Resolution? We shall discuss this next.

3.3.1 Resolution

Those who do not know much about microscopes ask, or boast, about magnification; those who know something about the optics of the microscope talk about resolution. The real quality of a lens, and in our case of a microscope, is its capacity to form a clear image of two very closely adjacent points. This is limited by the wavelength of the light being used: if two points are separated by less than half the wavelength of the light, they cannot be resolved, that is they cannot be reproduced as a sharp image showing two points. The wavelength of green light is about 550 nm (about half a micrometer), and thus points less than 250 nm apart cannot be resolved as separate points no matter how good the lens system. This of course suggests the great advantage offered by the electron microscope: the wavelength of the electron beam is tiny in comparison with that of light, and thus much greater resolution is possible. One might feel that working with blue light (with a shorter wavelength than green) would be helpful. In practice it doesn't make much difference. Ultraviolet light can be used but, unfortunately, the eye is not sensitive to this wavelength and photographic techniques are needed to record the images produced. Special quartz lenses are also needed.

Although 250 nm or so is the limit of resolution, we need not worry too much about this. What limits the actual resolution produced by lenses is not the wavelength of the light, but the capacity of the lens to capture light that has passed through the specimen. This is why working with blue light does not make much difference to the resolution produced. When light passes through an object made up of closely placed structures, the light is diffracted by the edges of those structures. A diffraction grating is the device used by physicists to study diffraction: the lines of the grating are very close together. We might simplify the argument by saying that some rays are not diffracted, some are diffracted to a given extent (bent away from the straight-through ray to a given extent) and some are diffracted to twice that extent. We might think in terms of the first and second diffracted rays. Abbe, who introduced the apochromatic objective, developed the theory of image formation based on the capture of these diffracted rays. The image is formed by interaction of the "straight-through" undiffracted ray with the diffracted rays, each being subject to refraction by the lens. Note that two effects are now in play: diffraction caused by the fine structure of the specimen and refraction caused by the lens. The capacity of the lens to distinguish two closely placed points depends, critically, on the capacity of the system to capture the diffracted rays.

Consider a point on a section. Diffracted light fans out from the point in three dimensions and forms a hemisphere. The resolving power of the lens will depend on how much of this hemisphere of

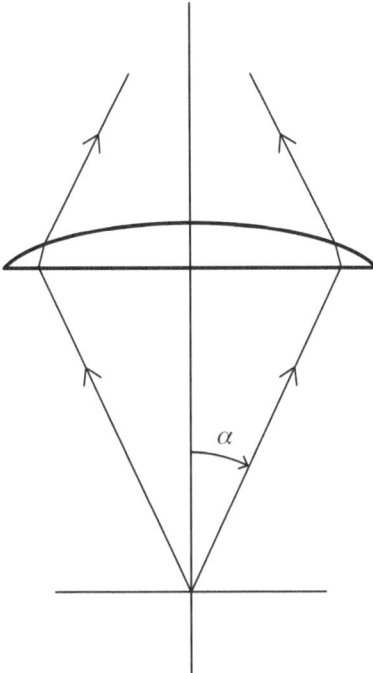

Figure 3.3 **The objective lens.** Path of light rays through an objective lens showing the acceptance angle and the half angle α.

light can enter the lens and be refracted by it. If only a part of the hemisphere can be collected, we can think in terms of collection of a cone of light. The wider the cone that can be accepted by the lens then the better will be the resolution as more of the diffracted rays will be used to form the image. The capacity of the lens to capture the cone of light is described as its numerical aperture. If the cone is defined in terms of its half angle α (see Figure 3.3) then its numerical aperture (NA) is defined by eqn 3.1:

$$NA = n\sin\alpha \tag{3.1}$$

where n is the refractive index of the medium between the specimen and the lens.

The resolving power of the lens is defined by eqn 3.2:

$$r = 0.61\lambda/n\sin\alpha \tag{3.2a}$$

or

$$r = 0.61\lambda/NA \tag{3.2b}$$

where λ is the wavelength of the light and r is the minimum distance between two points (or thin lines) that allows clear resolution of both points (or lines).

This is the fundamental equation of light microscopy and is well worth thinking about in a little more detail. Let us look first at the refractive index term n. Let us assume that the object was adjacent to the objective lens: not separated from it by anything. Then, at the limit, as mathematicians say, the half angle of the cone would be equal to 90°. We have defined a hemisphere rather than a cone! Now imagine the object a little further from the lens but still in contact with the lens *via* some medium that has exactly the same refractive index as the lens. As the object moves further away, the half angle of the cone will be reduced and the resolving power of the system will also be reduced. Let's do the calculation for the limiting case.

The refractive index of glass is about 1.5, sin90 is 1, and let's take the wavelength of the light to be 550 nm. Thus, the limit of resolution would be $(0.61 \times 550)/1.5 = 223.7$ nm. Now, a perfectly acceptable immersion objective (immersion because we are not thinking, yet, about any air gaps in the system) might have an NA of 1.3. Let us put $n = 1.5$, then $\sin\alpha$ is $1.3/1.5 = 0.87$. Thus, α is 60°. This is some way away from our theoretical limit for α of 90°. What will the resolution be? It will be: $(0.61 \times 550)/1.3 = 258$ nm. If such resolving power were available, we might be happy indeed. Sadly, such power is not generally available. Note that reducing the numerical aperture to 0.75 reduces the resolution to 447 nm: a significant, but not huge reduction. A red blood cell has a diameter of about 7 µm: resolving the cell as something distinctly different from a blurred point would be possible with any of these lenses.

We stressed above that there were no air gaps in the system. This is the case if the immersion technique is used and if the refractive index of the oil used to fill the gap between the coverslip over the section and the objective is the same as that of glass. Now consider what will happen if there is an air gap present. Light leaving the section will travel upwards in a straight line until it meets the interface between the glass of the coverslip and the air. At this interface it will be refracted away from its straight line. Consider a ray of light meeting the interface at 90°. It will be bent away from 90° by refraction at the interface and will leave the interface travelling outwards at an angle. Rays meeting the interface at an angle will be bent away similarly and those rays that meet the interface very obliquely will not get through it at all: they will be reflected back from the surface (see Figure 3.4). Thus refraction at the glass–air interface limits the angle of the cone

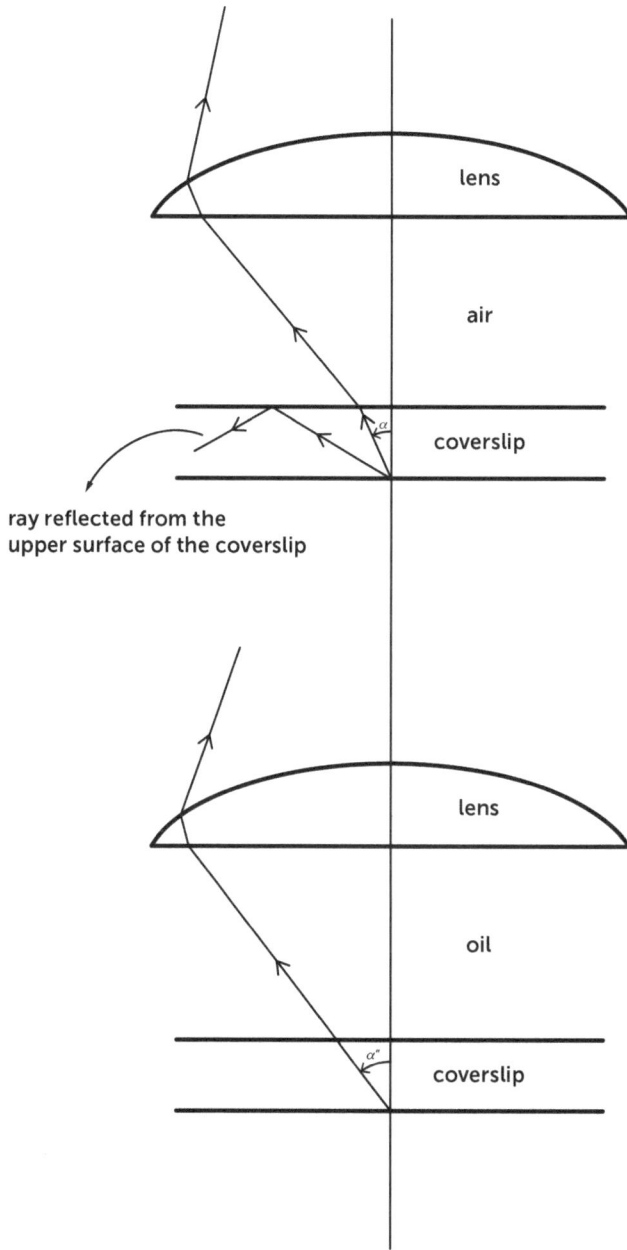

ray reflected from the
upper surface of the coverslip

Figure 3.4 **Effect of oil immersion on numerical aperture.** The widths of the air
and oil gaps have been exaggerated for clarity.

of light that can enter the objective. The air gap sets the limit to the size of the cone and the formula for the resolving power of the lens reflects this by replacing $n = 1.5$ by $n = 1.0$. The refractive index of air is, of course, 1.0. The numerical aperture of the system is defined by the lowest refractive index of the media that make up the system. If there is an air gap then the NA is limited to 1.0.

Now then, if this is the case, the resolving power of the system is reduced. Even at the limit, when the half angle of the cone in air is $90°$, the NA cannot exceed 1.0. Our limit for the resolving power of the lens will be: $(0.61 \times 550)/1 = 335.5$ nm. A good quality, dry (because we are not using it with immersion oil) apochromat that has a magnification of $20\times$ might have a NA of 0.6. The limit of resolution will be $(0.61 \times 550)/0.6 = 559$ nm.

The lesson from all this theory is that the resolving power of the objective is dependent on its numerical aperture and no lens working in air can equal the resolving power of a lens working in a satisfactory immersion medium. Lenses designed for immersion in water (refractive index: 1.333) and a range of other fluids have been designed. These are not widely used today, although a case for the water immersion system can be made: water is easy to use!

3.3.2   Magnification

The object is magnified to make it visible. Magnification of the object, per se, does nothing to improve the resolution: that was defined by NA. The object should be magnified so that such detail that has been made available by resolution can be seen clearly. Deciding how much magnification is needed involves knowing something of the resolving power of the eye. The Snellen eye chart is based on the assumption that people with average eyesight can resolve points separated by 1 minute of arc. Some people can do a little better than this.

If we imagine viewing these two points from a distance of 25 cm then, from the diagram (shown in Figure 3.5) and noting that:

$$0.5 \text{ minutes of arc} = 0.5/60 = 0.0083°, \qquad (3.3)$$

we can see that the distance between the points is given by:

$$2(25\tan0.0083) = 0.0072 \text{ cm} = 0.072 \text{ mm} \qquad (3.4)$$

A little more magnification than that needed to produce a separation of points of 0.072 mm is useful to most people and a rule of thumb has long been established. This states that magnification should not exceed 1000 times the NA of the objective. More

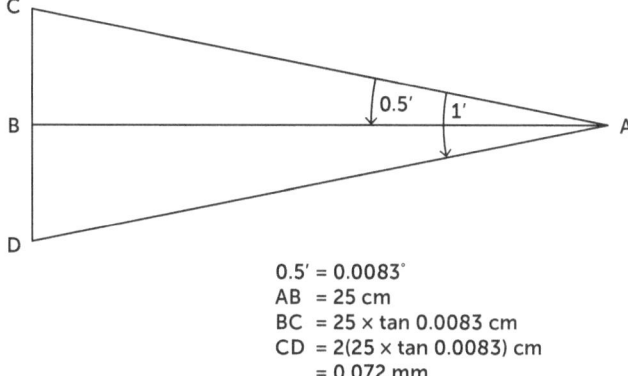

$0.5' = 0.0083°$
AB = 25 cm
BC = 25 × tan 0.0083 cm
CD = 2(25 × tan 0.0083) cm
     = 0.072 mm

Figure 3.5   Calculation of limit of resolution of normal eye-sight.

Table 3.1   Necessary magnification.

A	B	C	D	E	F	G
NA of objective	Resolution (μm). Points separated by the distances shown are resolved	Magnification needed to increase distance between just-resolved points to 0.072 mm	Magnification provided by objective	Total magnification, assuming the eyepiece provides 10× magnification	Barer's "minimum magnification" based on 250×NA	1000× NA
0.16	1.72	41	4	40	40	160
0.32	0.86	84	10	100	80	320
0.65	0.42	171	25	250	163	650
0.95	0.29	248	40	400	238	950
1.40	0.20	360	100	1000	350	1400

magnification than this will not enable us to see any more detail. Note that this is the maximum "permissible" or "useful" magnification. The analogy with a newspaper photograph is often mentioned. The detail is defined by the size of the dots used during printing. If such a photograph is examined with the naked eye at a distance of 20 yards, then not a lot of the detail will be seen: the eye lacks the necessary resolving power to make clear the details of the picture. If the picture is examined at 25 cm, then all the available detail will be seen. Magnification with a reading glass will not show any more detail: the individual dots will, of course, become visible.

The magnification produced by the combination of the objective and the eyepiece is, rather obviously, the magnification produced by the first multiplied by the magnification produced by the second.

Let us look at a table (Table 3.1).

Compare the figures in column C with those in columns E, F and G. The levels of magnification shown in column C are those that would allow two points separated by the distances shown in column B to be seen by somebody with very sharp eyesight. A little more magnification than this would be useful to many people: column E shows that this can be provided by combining a 10× eyepiece with the objective. Barer (see Bibliography) suggested that the magnifications shown in column F, which are very close to those in column C and calculated on the basis of 250×NA, as a guide to the minimum total magnification necessary. Column G shows the magnifications provided on the basis of 1000×NA: it will be seen that these far exceed the minimum necessary levels of magnification. Keeping the total magnification between the minimum calculated as 250×NA and the maximum calculated as 1000×NA is sensible and combinations of eyepieces and objectives used should be chosen on this basis. Another reason for keeping the total magnification above the minimum necessary (250×NA) will be discussed later when we consider the "exit pupil" of the eyepiece.

The magnification produced by the objective is also defined by the length of the tube of the microscope. This is reflected in eqn 3.5:

$$\text{Magnification} = \text{tube length/focal length} \tag{3.5}$$

The usual tube length is 160 mm. Thus, a 10× objective will have a focal length of 16 mm and a 40× objective will have a focal length of 4 mm.

3.3.3 Focal Length, Working Distance, Diameter of Field and Depth of Field

Objectives of high numerical aperture and high magnification have short focal lengths. Recalling that the object is placed just outside the primary focus of the objective it will be clear that the working distance between the front of the objective and the top of the coverslip will also be limited when high-power objectives are used. Table 3.2 provides a list of focal lengths and working distances for a range of objectives.

Note that with high power, high NA immersion objectives the working distance is very short indeed: not much more than a tenth of a millimetre. Note that even with high power dry lenses the working distance is much less than 1 mm; this explains why putting the slide upside down on the stage of the microscope will defeat your efforts to focus the section.

Table 3.2 Focal lengths and working distances.

Objective type	Magnification	NA	Focal length (mm)	Working distance (mm)
Achromat	10	0.25	16	7.70
Apochromat	10	0.30	16	4.85
Achromat	20	0.50	8	1.60
Apochromat	20	0.65	8.3	0.50
Achromat	45	0.85	4	0.30
Apochromat	47.5	0.95	4	0.18
Achromat	97 oil	1.25	1.8	0.13
Apochromat	90 oil	1.30	2.0	0.12

Table 3.3 Depth of field.

NA	0.25	0.30	0.50	0.65	0.85	0.95
Depth of field (μm)	8.52	5.83	1.91	0.99	0.40	0.19

The diameter of the field of view is also reduced as NA and magnification are increased. For the 10× lenses listed in Table 3.2, the diameter of the field is about 1–2 mm; for oil immersion lenses, the diameter of the field of view will be only 0.1–0.2 mm.

3.3.3.1 Depth of Field. More interesting perhaps than the diameter of the field of view is the depth of field. Consider a histological section 15 μm thick. The microscope could be focused on the top surface of the section or any optical plane through the section down to its bottom surface. Unsurprisingly, objects at different levels will not be all in sharp focus at the same time: those above and below the plane of focus will be slightly out of focus. The depth of field of an objective specifies the vertical distance within which objects will all appear reasonably in focus. Table 3.3 gives the depths of focus for a range of numerical apertures.

The depths of field specified in Table 3.3 are rather less than those often quoted. This is because the figures are for photographic depth of field. This is calculated as shown in eqn 3.6:

$$d = \frac{\lambda\sqrt{(n^2) - (\text{NA})^2}}{(\text{NA})^2}, \tag{3.6}$$

where λ is the wavelength of the light, n is the refractive index of the medium and NA is the numerical aperture. The key message is that low NA means large depth of field and high NA means low depth of field. This is important in photographic work: all the out of focus objects will scatter and diffuse the light passing through the specimen and produce blurring of the image.

Depth of field is less of a problem for purely visual examination of sections. This is because the eye can accommodate for the depth of objects within the section. In addition, of course, the microscopist is constantly adjusting the fine focus of the microscope and is effectively focusing up and down through the depth of the section. In practice the visual depth of field (VDF) in millimetres is given by eqn 3.7.

$$VDF = 250/M^2 \tag{3.7}$$

where M is the total magnification of the microscope set up.

Thus, with a 10× objective and a 10× eyepiece, VDF = 250/10 000 = 0.025 mm = 250 µm.

And, with a 100× objective and a 10× eyepiece, VDF = 250/1 000 000 = 0.00025 mm = 0.25 µm.

This means that, at low power and once the section is in focus, you do not need to focus up and down with the fine focus control. It also means that, when working at high power, constant adjustment of the fine focus control is essential. Depth of field can be increased, at the expense of resolving power, by stopping down a high-NA objective by means of the aperture (sub-stage) diaphragm. The proper use of this diaphragm will be discussed in Section 3.4 dealing with the condenser.

3.3.4 Depth of Focus

Users of microscopes often become confused by the terms "depth of field" and "depth of focus". Depth of field refers to the plane of the section. Depth of focus refers to the plane of the photographic film used in photo-microscopy.

Depth of focus is defined by eqn 3.8:

$$\text{Depth of focus} = \text{depth of field} \times M^2, \tag{3.8}$$

where M is the total magnification provided by the microscope.

Shillaber (see Bibliography) gave the figures in Table 3.4.

Table 3.5, modified from the Nikon website, provides useful figures.

Depth of focus has been calculated using a different formula from that given in eqn 3.8. However, it will be seen that low-NA, low-magnification objectives have lower, not higher, depths of focus than high-NA, high-magnification objectives. This is important in

Table 3.4 Depth of field and depth of focus.

Focal length (mm)	Total magnification	NA	Wavelength (μm)	Depth of field (μm)	Depth of focus (mm)
16 mm	300	0.3	555	5.88	529.2

Table 3.5 Depth of field and depth of focus.

Magnification	NA	Depth of field (μm)	Depth of focus (mm)
4×	0.10	55.5	0.13
10×	0.25	8.5	0.80
20×	0.40	5.8	3.8
40×	0.65	1.0	12.8
60×	0.85	0.40	29.8
100×	0.95	0.19	80.0

photo-microscopy. Slight errors of focusing make little difference to the sharpness of the photographs taken with high-NA objectives, but when low-NA objectives are used, even a slight error in focusing is reflected in a very out of focus print. Anybody who has done some photo-microscopy knows that the real test of focusing is to get your low-power prints in sharp focus. Getting the high-power prints in sharp focus is a lot easier.

3.3.5 Coverslips and Spherical Aberration

All good-quality objectives are corrected for spherical aberration. When the lens is designed, the calculations involve taking into account the thickness of the coverslip used to cover the histological section. The coverslip is assumed to be of a given thickness, for example, between 0.16 and 0.19 mm. If a coverslip thicker or thinner than was allowed for in the calculations is used then spherical aberration will be introduced. Of course, the thickness of the mounting medium between the section and the coverslip should also be considered. The layer of mounting medium is assumed to be as thin as possible. As noted above, spherical aberration makes it impossible to produce a perfectly sharp image. This is especially a problem when using high-power dry (dry means un-immersed) objectives. If the coverslip has an important effect, it should be obvious that we could eliminate this effect by using an immersion system. This is so as long as the refractive index of the immersion fluid is identical to that of the coverslip and the objective. Spherical aberration can, however, be corrected for without the use of immersion. The easiest way to do this is to vary the length of the tube of the microscope. The calculations made when the lens was designed assumed a standard tube length. What this does is to introduce some spherical aberration in what might be called the opposite direction to that introduced by the coverslip. This simple method is, unfortunately, not available to the users of modern microscopes, where the tube length is fixed; it was available on older, monocular, instruments! High-power immersion objectives require thin coverslips: check the thickness of the ones you use.

A second method is for the manufacturer to provide a correction collar on the objective. This is a calibrated ring that can be rotated. Rotating the ring changes the correction of the lens for spherical aberration. The use of the correction collar will be discussed later but note that such adjustment is not, in practice, needed with objectives of numerical aperture less than 0.75.

Keeping the layer of mounting medium as thin as possible is important. Layers that are too thick are often produced by applying the mountant to the section and then laying the coverslip onto the mountant (see Chapter 7 for details of mounting sections) because the coverslip lacks the weight necessary to press the mountant out of the way. This may be a problem if the mounting medium is too thick: thinning the mountant with a suitable solvent, for example, xylene, helps. Placing a small weight on top of the coverslip to press it down onto the section is also helpful. Inverting the process—putting the mountant on an upside down coverslip and then laying the slide carrying the section onto the mountant—helps because the slide is heavier than the coverslip and presses out the mountant rather more effectively. Even better is mounting the section on the under-surface of the coverslip and not on the slide; although this introduces difficulties during staining in that delicate coverslips rather than comparatively robust slides have to handled.

None of these methods is as effective in practice as using an immersion system if the immersion medium has the same refractive index as the glass of the coverslip.

3.3.6 Labelling of Objectives

All modern objectives carry a wealth of information inscribed on their barrels, as shown in Figure 3.1. The information usually includes:

- The numerical aperture of the lens;
- The magnification of the lens;
- Whether the lens is an achromat or an apochromat.

Correction for chromatic aberration involves the use of fluorite lenses. Lenses that offer correction for chromatic aberration somewhere between that provided by achromats and proper apochromats are known as fluorite objectives or semi-apochromats. Fluorite objectives are really very good and significantly less expensive than "full apochromats".

If the lens has been corrected so as to produce a flat field, the word "plan" (for plane) is often included. An apochromatic lens corrected so as to produce a flat field is known as a "plan-apo". The flatness of the field can easily be judged by looking at a section. If the edges of the field are a little blurred when the centre of the field is in sharp focus, then the lens is not producing a flat field. Some very fine, older apochromatic lenses produced splendid resolution despite the fact that they did not provide a flat field.

The thickness of the coverslip for which the lens has been constructed might also be provided.

The tube length for which the lens has been designed is also sometimes provided.

If the lens is designed for oil immersion, the word "oel" or "oil" is included.

All the essential data are, for example, provided on an objective produced by Leitz:

"160/0.17PL APO OEL 100/1.32".

Meaning:

Tube length: 160 mm;
Coverslip thickness for which the objective is adjusted: 0.17 mm;
PL: "plano" or flat field;
APO: apochromatic;
OEL: oil immersion;
100: magnification;
1.32: Numerical aperture.

A very desirable lens!

This is a lot of information: surprisingly few people seem to know that it is available. Much less information appears on older objectives.

3.3.7 Spring-loaded Objectives

The lower part of a modern, high-power objective is spring loaded. This is not generally the case with low-power objectives. The spring loading protects the front lens of the objective: if the lower surface of the objective accidentally comes into contact with the coverslip, it will retract rather than being driven through the coverslip if advanced any further. The short working distance of high-power objectives makes this a very useful feature! Note that the front lens of the objective cannot be far recessed in the mount because of the short working distance of the lens. Older high-power objectives were not spring loaded and have to be used with extra care, especially if they are not parfocal with the lower power objectives that are available on the microscope. Parfocality means that if one objective is in focus then all the others should also be in, or very close to being in, focus. When focusing down on a section with a high-power objective, you may notice that the fine focus control suddenly becomes a little stiffer to turn. The lens is in contact with the coverslip and the spring loading is being compressed. STOP AT ONCE! One of the great advantages of the parfocality of modern objectives is that if one objective is in focus the others will be almost in focus. Thus, only adjustment of the fine focus control should be necessary when adjusting a high-power objective. The golden rule, **begin with a low-power objective and work up to higher powers**, should always be observed.

3.4   THE CONDENSER

The condenser is the Cinderella of the microscope. Many users know something about objectives and specify high-quality lenses when ordering a new microscope; few pay much attention to the condenser. This is a great pity because the condenser controls the functioning of the objective and indeed of the microscope as a whole. Failure to produce a really first class image, or to attain the limit of resolution of which the objective is capable, is usually due to either the use of an inadequate condenser or to the poor adjustment of an adequate condenser.

It is possible to operate a microscope without a condenser: some instruments used in elementary biology classes have only a mirror that directs the light, from a lamp or from the window, up into the microscope. The mirror should be concave so as to focus the light on the section. A piece of ground glass placed under the section is useful in that it spreads out the light and allows a larger area to be illuminated.

All microscopes intended for serious work have a condenser. The condenser can work with a mirror: the mirror conducts the light to the condenser, which then focuses it on the section. On this sort of instrument the mirror usually has a plane side and a concave side. The plane side is for use with the condenser; the concave side is for use when the condenser has been removed, as might be the case when very low-power objectives are being used. The concave mirror acts as a condenser with a very low numerical aperture: it produces a narrow cone of light, which is satisfactory for low NA objectives but not for high NA objectives. The condensers fitted to microscopes intended for class use rather than for research (it being assumed that research workers require an instrument with better performance) are usually rather simple. The distinguished German physicist Ernst Abbe designed a simple condenser, the Abbe condenser, and it is still fitted to many microscopes intended for use by students. Abbe referred to this condenser as an illuminator. The Abbe condenser is not highly adjusted and should only be used when high resolution is not needed. High-quality condensers (Abbe also designed some of these) are designed in the same way as objectives: they are corrected for chromatic aberration and are designed to focus light in one plane. Thus, an achromatic-aplanatic condenser is fitted to microscopes intended for highest quality work. In the early days of the development of the modern microscope a great deal of attention was paid to condensers and some enthusiasts inverted objective lenses and used them as condensers. This is unnecessary.

What, precisely, does the condenser do? It does two things that are extremely important: it provides an evenly illuminated field and, most important, it focuses a cone of light on the section. This width of the cone should match that received by the objective. This is a key point. Ideally, the numerical aperture of the condenser should match that of the objective. Let us look at some more ray diagrams. The next section will require some concentration but making the effort is well worthwhile.

The diagrams are taken from *The Royal Microscopical Society Handbook* (01), "An introduction to the optical microscope" by Savile Bradbury (see Bibliography). It is fair to say that he who understands these diagrams understands how a microscope works. There is little to be gained by discussing the condenser without also discussing its main purpose: to provide light to the section. We shall therefore discuss the condenser in terms of illumination.

In Figure 3.6, the diagrams A and B show, respectively, the paths taken by the "image-forming" and the "illuminating" rays. This may cause some confusion: no actual separation between image-forming and illuminating rays exists. All that has been done is to employ a device that makes thinking about the system easier.

Look at the diagram of the illuminating rays (Fig. 3.6B). Note that these begin at a point on the filament of the lamp at the bottom of the

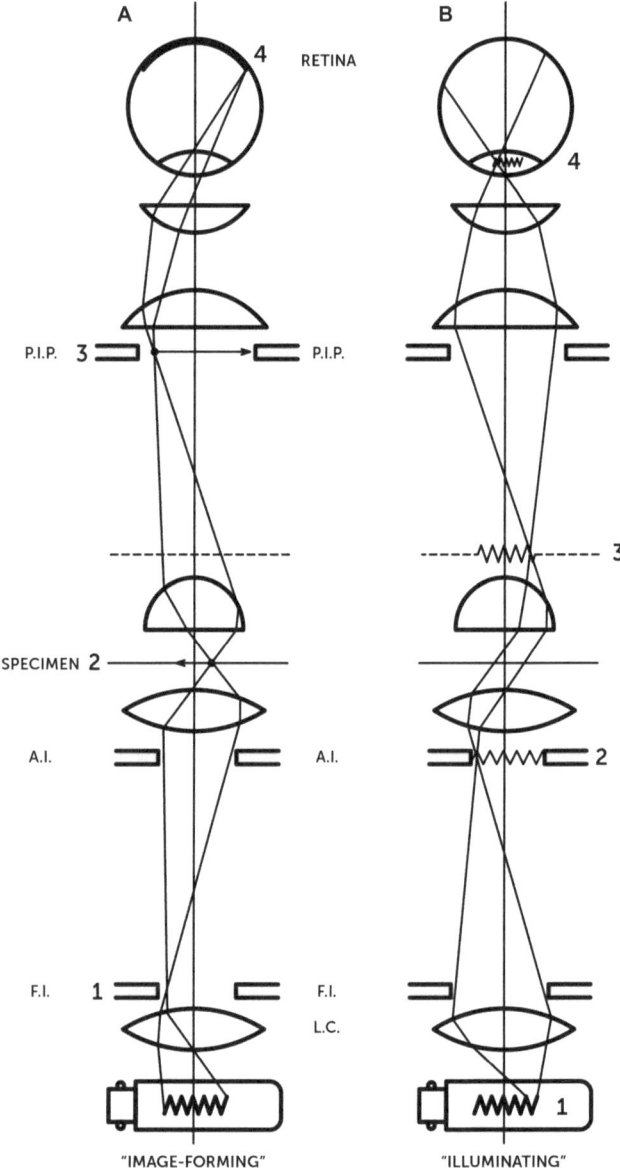

Figure 3.6 **Diagrams illustrating the "image-forming" (A) and "illuminating" (B) rays in a microscope adjusted for Köhler illumination.** LC: lamp collector lens, FI: field iris diaphragm, AI: aperture iris diaphragm or sub-stage diaphragm, PIP: primary image plane. Conjugate planes are indicated by the numerals 1, 2, 3 and 4 in each diagram.
Modified from Figure 14 in: "An Introduction to the Optical Microscope", S Bradbury 1984. Diagrams reproduced by permission of the Royal Microscopical Society.

diagram. All the rays from this one point are brought to a focus at the aperture diaphragm just below the sub-stage condenser. Only the rays from one point on the filament have been drawn in. Imagine that we drew in rays from other points on the filament: these too would be brought to a focus at the aperture diaphragm. In fact a perfect image of the lamp filament would be produced at the aperture diaphragm. Because the image of the filament is formed at the lower focal plane of the sub-stage condenser, rays entering that lens will leave it as a parallel beam. These will pass through the plane of the section and be brought to a focus by the objective at the back focal plane of the objective. The width of the beam of parallel rays can be controlled by the diaphragm in front of the lamp condenser lens: this diaphragm is known as the field diaphragm or as the illuminated field diaphragm.

Now let's turn to the image-forming rays (Figure 3.6A). We are thinking about image formation in terms of the image of the section and not, of course, about the images of the lamp filament, which we have just discussed. Let us imagine that these image-forming rays come from the lamp condenser lens. In the diagram these rays are shown as starting from the plane of the field diaphragm, just in front of the lamp condenser lens. Note that these rays diverge as they pass upwards and are focused by the sub-stage condenser onto the section. If we put some object in the plane of the field diaphragm, say a wire grid, we would see its image in the plane of the section. These rays interact with the specimen, they are diffracted by the fine structure of the specimen, and light from the section is collected by the objective and focused as the primary image of the microscope. We imagined a moment ago that the image-forming rays began at the illuminated field diaphragm. Of course, they actually began at the filament of the lamp. This is shown in the diagram. This raises an important point: light from every part of the filament, from the whole of the filament, is focused on each point of the specimen.

3.4.1 Conjugate Planes

Now then, let's put it together. To do this, we need to think in terms of those terrifying things: "the conjugate planes". This phrase means, in general terms, that if something, for example, the lamp filament or the field diaphragm or the object being examined, is in focus at two or more different planes in the optical path, these planes are called **the conjugate planes for that object**. From the above description of the illuminating rays and the image-forming rays, we can see that there will be a number of conjugate planes.

3.4.1.1   Conjugate Planes for the Illuminating Rays.   Let us first look at the conjugate planes for the filament. These lie along the pathway we described for the illuminating rays. Starting with the filament, the first conjugate plane is at the aperture diaphragm of the sub-stage condenser. If we put a screen here, we would see, on its lower surface, an image of the filament. This is done in setting up the microscope for Köhler illumination. This need not bother us for the moment, but when we come to it a lot of what is currently being explained will be revised. The second conjugate plane will be at the back focal plane of the objective. The third conjugate plane will be just above the eyepiece. We can see the second conjugate plane by removing the eyepiece and looking down the tube of the microscope. The image of the filament is not all that easy to see: a rather special little telescope used when doing phase contrast microscopy can be used to see it if you have such a device to hand! The third can be seen by putting a piece of thin paper above the eyepiece and looking down at it. You won't see the filament, its image is too small at this point; what you will see is a small circle of light, but if you use a $10\times$ magnifier instead of the card you will see the filament.

3.4.1.2   Conjugate Planes for the Image-forming Rays.   Let us begin at the field diaphragm. An image of the field diaphragm is formed in the focal plane of the section. You can see this very easily by looking at the specimen in the usual way. Open and close the field diaphragm whilst looking at the specimen. If the field diaphragm looks blurred, adjust the sub-stage condenser to bring it into sharp focus. When it is in sharp focus, we can honestly say that the field diaphragm has been imaged in the plane of the section. The next conjugate plane is at the primary image of the object. What is being formed in this plane is an image of the specimen and an image of the field diaphragm. The final conjugate plane is, of course, on your retina, where the image of the section finally appears. Open and close the field diaphragm and notice the change in the area of the section that is illuminated.

So, we have a series of conjugate planes for the filament and another series for the field diaphragm. To the latter series is added the image of the section. We can think of illuminating rays beginning at a point on the filament of the lamp and image-forming rays beginning at a point in the plane of the field diaphragm. But, of course, all this has been but a device to help us understand how the microscope works. In reality no rays can be labelled illuminating rays and no rays can be labelled image-forming rays.

3.4.2 Illumination of the Specimen and the Functions of the Lamp Condenser Diaphragm and the Sub-stage Diaphragm

This section repeats a good deal of what has been said already. This is deliberate. The current section takes a more practical approach than the earlier discussion and is designed to help in setting up the microscope.

The production of a satisfactory, detailed image of a histological section depends on the section being illuminated brightly and evenly. A number of methods of ensuring this have been developed, but perhaps the best is the method referred to as Köhler illumination. This method was developed by August Köhler in the 19th century. He worked with an optical bench and was especially interested in creating a form of illumination that was particularly suitable for photomicrography. For such work, completely even illumination was necessary and it was important that the light source (a gas flame or the filament of a lamp) should not appear in the photograph. He thus decided that methods that involved throwing an image of the illuminator (the flame or the filament) into the same plane as the section were unsatisfactory. It might be noted, in passing, that such methods were referred to as providing "source-focused" or "critical" illumination and were much used by those interested in achieving maximum resolution.

Now for an experiment. This has been adapted from a description provided by Evennett and by Barer.[†] Light a candle and stand it on top of a four-drawer filing cabinet, or some other suitable piece of furniture. You will now need a reading glass and a piece of white paper or card. The author uses a 9 cm diameter lens (a large reading glass) and a focal length of about 19 cm. Hold the card about a yard away from the candle flame. Hold the lens close to the flame and move it towards the card. As you move the lens away from the flame, an inverted and enlarged image of the flame will appear on the card. Note how far away from the flame the lens is when the image is sharp. Now replace the card with your eye and continue to hold the lens where it was when the enlarged, inverted image of the flame was produced. Look through the lens at the flame. Move the lens towards and away from the flame very slowly. You will see the large, inverted flame if the lens is a little too far away, and a large upright flame if the lens is a little too close to the flame; however, at one point, when the flame is focused on your eye, the entire lens will glow with light and no flame will be visible. This is the critical point. This is precisely what is happening with Köhler illumination. The reading glass is playing the

---

[†]Barer: see bibliography. Evennett P: *Proc. R. Microscop. Soc.*, 1994; 28: 10–13.

part of the lamp condenser lens. The lens of the sub-stage condenser collects light from the brilliantly illuminated lamp condenser lens and focuses it onto the specimen. The beauty of the method is that light from this secondary source, the brilliantly lit lens of the lamp condenser, is focused on the specimen. The brilliantly illuminated lens of the field condenser acts as a secondary light source. Barer pointed out that the sub-stage condenser lens "sees" the lamp condenser lens as a brilliantly illuminated circular area of light.

3.4.2.1 *Back to Köhler.* Köhler arranged for a lens placed in front of the filament of the lamp (the lamp condenser lens) to throw an image of the filament towards a second lens placed in front of the specimen on the stage of the microscope, which was arranged horizontally (the tube of the microscope was horizontal and the stage was vertical) on his optical bench. The second lens is the lens of the sub-stage condenser. The lamp condenser lens was arranged so that the image of the filament lay precisely in the lower focal plane (we might call it the first focal plane) of the lens of the sub-stage condenser, the second lens in his set up. Light from the image of the filament passed through the second lens and emerged as a series of parallel (or nearly parallel) rays: the section was evenly illuminated by these rays.

In this simple system the sub-stage condenser could have been fixed in place: it would need no adjustment in that all it is required to do is to collect rays from the image of the filament and transmit these as parallel rays through the specimen. A refinement was then developed.

It was noted that limiting the area illuminated to the area that could be seen *via* the objective lens of the microscope improved the quality of the final image viewed *via* the eyepiece of the system or projected by the eyepiece on to photographic film by reducing reflections from inside the microscope tube. How could this be done?

The answer was to place a diaphragm just in front of the lamp condenser lens. To control the area of the specimen that is illuminated, this diaphragm (the field diaphragm, better described as the illuminated field diaphragm) needs to be seen clearly in the plane of the section. How could this be done? Very easily indeed by (1) focusing the specimen and (2) adjusting the position of the sub-stage condenser lens so that an image of the field diaphragm is projected into the same plane as the section. This was done before the lamp condenser lens was adjusted to throw an image of the filament to the focal plane of the lens of the sub-stage condenser. Note that the lamp condenser lens can be adjusted, but that the field diaphragm is fixed in place. We are now in a position to understand the steps taken in setting up Köhler illumination.

1. Focus the section.
2. Focus the field diaphragm in the plane of the section by adjusting the sub-stage condenser. When the field diaphragm is imaged in the plane of the section, it may be noticed that it is off centre. The sub-stage condenser should be adjusted by means of the centring screws. As these are turned, the image of the field condenser will move about in the plane of the section and will be visible via the eyepiece. Adjust the centring screws until the circle of light expands evenly to fill the field of view when the field diaphragm is opened.
3. Adjust the lamp condenser so that an image of the filament is thrown by the lens of the lamp condenser to the lower focal plane of the sub-stage condenser lens.

A problem has now appeared. We can see the field diaphragm by looking down the microscope in the usual way and we can put it into sharp focus, but how can we know that the lamp condenser is actually throwing a really sharp image on the filament into the lower focal plane of the sub-stage condenser? We need to know this so that step 3, above, can be completed. We might begin by saying: if it is doing this then the specimen should be evenly illuminated by the parallel, or nearly parallel, rays coming from the sub-stage condenser lens; the image of the filament being, then, in the lower focal plane of the sub-stage condenser. We might judge this by looking at the image of the section and deciding whether it was, in fact, evenly illuminated. Our ability to judge might, however, not be all that good. A better, though slightly awkward, method is to bend down and hold a piece of white card just below the sub-stage condenser (putting a circle of card into a filter holder, which is often available just here, is even better). If the lamp condenser is throwing a sharp image of the filament onto the card we might be reasonably happy. The card could, of course, be viewed with a small mirror to avoid bending down and looking up at the sub-stage condenser. But, better still, we can examine the back focal plane of the objective lens by removing the eyepiece and looking down the tube. The filament will be visible, though seeing it takes a little practice; as described above, a phase telescope could be used. A very good method is to use a magnifying glass placed just above the eyepiece lens: here too the filament will be visible. It will be clear that the filament is in focus at a number of positions along the optical path of the microscope. This is illustrated in Figure 3.5.

Setting up Köhler illumination is easy as long as all the components mentioned above are available and can be adjusted according to the plan. Sadly, this is not always the case. Modern microscopes are often designed to be less, rather than more, adjustable than older instruments. A ground glass disc is sometimes inserted by the manufacturer into the optical path, often as a part of the illuminator. This prevents the filament being seen. Sometimes the ground glass disc can be swung out of the way and Köhler illumination can be set up properly. In better instruments the ground glass disc is so positioned that it becomes a substitute for the filament: it becomes, effectively, the primary light source. If this is the case and if the disc cannot be swung out of the way, then all we can do is to ensure that the field diaphragm (this is always present and adjustable in the sense that it can be opened and closed) is focused in the plane of the section by adjusting the sub-stage condenser, and opened so that the illumination just fills the area that can be observed. One might wonder why the ground glass disc is included. With low-power objectives it is sometimes impossible to set up Köhler illumination in such a way that the filament is invisible. This is due to the great depth of field of low-power objectives. This is not a problem with high-power objectives (small depth of field) and some manufacturers recommend that the ground glass disc is swung out of the way when high-power objectives are used. Microscopes with built-in illumination systems are, too often, not provided with any means of adjusting the lamp condenser. It may be impossible to focus the filament in the lower focal plane of the sub-stage condenser. To do this we need to be able to adjust the position of the lens of the lamp condenser. Many microscopes with built-in illumination do not allow this. If this is the case, we must trust that the manufacturer has done this for us. All that can be done is to look for the image of the filament in the conjugate planes described above and to complain to the manufacturer if it cannot be seen or if it appears out of focus. Before complaining it is as well to check that no ground glass screen has been included in the optical pathway because this will prevent you seeing the filament!

3.4.2.2 The Sub-stage Condenser Diaphragm.   It will have been noticed that nothing has yet been said about the diaphragm of the sub-stage condenser. This, by the way, should have been opened as widely as possible before starting to set up the instrument for Köhler illumination. What does it do? When and how should it be adjusted? This diaphragm controls the angle of the cone of light that we might imagine impinging on any point in the section. The rays that make up this cone are the image-forming rays coming from the brilliantly illuminated lens of the lamp condenser, and optimal image formation occurs when the angle cone of illuminating rays precisely matches the angle of the cone of light that enters the objective lens. We might think of two cones: one standing on its base, the other balanced on top of it—tip to tip. This is achieved by closing the sub-stage diaphragm. By adjusting the sub-stage diaphragm, we are actually adjusting the numerical aperture (already discussed) of the sub-stage condenser and trying to match it to that of the objective lens. How do we know when it is properly adjusted? There are two methods. If the eyepiece is removed and we look down the tube and, at the same time, adjust the sub-stage diaphragm, we will see it narrowing and widening. Set it so that two thirds to three quarters of the disc you can see is visible; some authorities say that nine tenths of the disc should be visible, but not many people do this. What most do is close the sub-stage diaphragm until the image gains in contrast. Begin with the sub-stage diaphragm wide open and observe the section. Close the diaphragm slowly: notice that the image becomes a little darker and that the edges of structures become more sharply defined. Keep on closing the diaphragm: the image will become very "contrasty" and dark. You have now closed the diaphragm too far! It should be noted that, as the image gains in contrast (and depth of field) at the expense of resolution, the edges of features become thickened and run into each other. Open it again and close it slowly until the image gains in contrast but not excessively. Microscopists often say that the sub-stage diaphragm (better defined as the illuminating aperture diaphragm) should be closed until the image acquires "oomph". Judging this requires practice. This diaphragm should never be used to adjust the brightness of the image: that is done by adjusting the rheostat of the illuminator. Sometimes this is an unsatisfactory approach to controlling brightness. Colour photography requires the lamp to be running at the correct "colour temperature" to produce an appropriate set of wavelengths: "turning down the wick" (*i.e.*, turning down the rheostat) ruins this. What can be done? The answer is to place a neutral density filter in the optical path. This will reduce the brightness, but will not upset the colour temperature. The neutral density filter can be put anywhere in the optical path: filter carriers are often provided as part of the sub-stage condenser or as part of the lamp condenser. In passing, it is useful to note that a pale blue filter produces a better (less reddish) coloured image for ordinary viewing: removing the blue filter is sometimes necessary for colour photography depending on the colour temperature of the lamp and the spectral sensitivity of the photographic film being used.

What, then, are the great benefits of Köhler illumination? There are in fact two. First, by placing an image of the light source, albeit an uneven light source, such as a filament of a lamp, at the lower focal plane of the sub-stage condenser, it allows a beam of parallel rays to illuminate the specimen. As Evennett has pointed out, the light source itself could, in theory, have been put in this position: at the lower focal plane of the sub-stage condenser. Clearly, this would have been difficult when using a gas flame, but a lamp filament could have been placed here. Obviously, this might have caused problems with overheating of the sub-stage condenser and of the section. Köhler's step forward required development of the lamp condenser lens to throw the image of the lamp filament into the lower focal plane of the sub-stage condenser. The second, and perhaps greater, step forward was in noticing that when the lamp condenser was arranged like this its lens glowed with light. When focused in the plane of the specimen, this glowing circular area of light provides an excellent secondary light source. Secondary? Yes, the filament is the primary source, but we know that it is not imaged in the plane of the section; the glowing lens of the lamp condenser is the secondary light source. Adding the field diaphragm just above the lamp condenser lens, noting that it could be imaged in the plane of the section and that it controlled the area of illumination completed Köhler's brilliant invention.

Examination of Figure 3.6 will show that rays arising from **each point of the filament** pass through **all parts of the section** and that **each point of the section** receives light from **all parts of the filament**.

Köhler in 24 words

Kohler illumination provides a brilliant, secondary light source and moves the image of the lamp filament away from the focal plane of the object.

3.4.2.3   A Reminder About the Diaphragms.   The lamp condenser diaphragm (the field diaphragm or the illuminated field diaphragm) controls the area of the section that is illuminated. Adjusting it controls glare which reduces the quality of the image: the reduction in glare improves contrast. Adjusting it does nothing to the NA of the system.

The sub-stage condenser diaphragm (the aperture diaphragm or the illuminating aperture diaphragm) controls the angle of the cone of rays that interact with the specimen and form the image. Adjusting this diaphragm controls the NA of the system, the contrast of the image and the depth of field. As this diaphragm is closed, the image becomes less bright but you should never, ever use this diaphragm to control the brightness of the image. For that, adjust the rheostat or add neutral density filters.

3.4.3 The Condenser, Again

Having exhausted ourselves in discussing illumination, let us return to the condenser itself. An air gap exists between the top of the condenser and the lower surface of the microscope slide. This air gap inevitably limits the diameter or angle of the cone of rays that pass upwards from the condenser. At the upper surface of the coverslip, rays of light will be refracted as they pass from glass to air and the more oblique rays will be reflected back towards the condenser. The numerical aperture of the condenser, just like that of the objective, is limited by the refractive index of the medium between the upper surface of the condenser and the lower surface of the microscope slide. The answer to this is to fill the gap with immersion oil. Only by doing this can the numerical aperture of the condenser be raised above 1.0. This is rather important if an oil immersion objective is being used. The golden rule is: *if there is an air gap in the pathway from condenser to objective, the effective numerical aperture of the objective is limited to 1.0.* Despite this, nearly all histologists use oil immersion objectives without immersing the top lens of the condenser. There are perfectly good reasons for doing so: a less than optimal numerical aperture is accepted in return for the elimination of spherical aberration introduced by the coverslip and mountant and an increase in numerical aperture over that, which might be produced by the highest power, dry objective, becomes available. The effects of the coverslip could, of course, be eliminated by soaking off the coverslip with xylene. Let us assume that your microscope has a 40×, 0.65 NA dry objective without a correction collar. This is, in fact, the 40× lens that the author has fitted to his Wild M20-EB microscope. Using an oil immersion 100×, 1.3 NA lens, even without immersion of the condenser (putting oil between the top lens of the sub-stage condenser and the under surface of the slide), will offer a significant improvement in resolution over that provided by the 40× lens. The oil immersion lens will be working with a NA of about 1.0 and spherical aberration will have been reduced. Of course, for optimal use of oil immersion objectives, the condenser must be oiled to the microscope slide. Oil should not be applied to a condenser unless it is designed for immersion. The NA of a condenser should be inscribed upon it by the manufacturer, who should add "oel" or "oil" to the labelling inscribed on the condenser if it is meant for use by oil immersion.

Condensers of high NA will not be satisfactory for use with low-power objectives. Such objectives have such a wide field of view that a high-NA condenser will not be able to fill the field with light. Many manufacturers provide a "two in one" condenser system with a top lens that may be tilted into and out of the optical path as necessary. These are described as "flip-top" condensers.

We know that NA controls resolution; we also know that the NA of both the objective and the condenser need to be taken into account when assessing the resolving power of the microscope. Eqn 3.9 is often quoted:

$$r = 0.61\lambda/((NA_{obj} + NA_{cond})/2) \tag{3.9}$$

This equation simply says that we should take the average of the NA of the objective and the condenser and use that in our usual calculation. Well, just about. As Barer has pointed out, if the condenser is simply removed, the resolving power does not fall to a half of what it was, although it is reduced. Similarly, if the objective is removed, the system will have no resolving power at all! The lesson is that the NA of the condenser should be similar to that of the objective, but they need not be identical. Whilst many microscopes offer a choice of four or more objectives, they are unlikely to be equipped with more than a choice of two numerical apertures for their sub-stage condenser. Microscopes in which the condenser assembly rotates with the objective carrier have been produced. Such instruments match the condensers with the objectives but seem not to have become very popular.

3.4.3.1 A Last Point About Condensers in General. Just as the objective is sensitive to the thickness of the coverslip, the condenser is sensitive to the thickness of the slide. A condenser with a high NA will have a short working distance and will not be able to provide a sharply focused image of the field diaphragm in the plane of the section: the image of the field diaphragm will, in fact, lie below the plane of the section. Slight unscrewing of the top lens of the condenser may allow the field diaphragm to be imaged in the plane of the section. Microscope manufacturers tend not to specify the optimum thickness of slides for use with their condensers: slides of 1 or 1.1 mm thickness are generally satisfactory.

3.4.3.2 Condensers for use with low power objectives. If very low-power objectives are in use then a standard condenser, however excellent it is with high-power objectives, might not fill the field of view with light. The remedy is to use a special condenser or to remove the condenser completely.

3.5 THE EYEPIECE(S)

The eyepiece magnifies the primary image and forms the virtual secondary image. In the description given above, the eyepiece was represented by a single biconvex lens, as indeed was the objective. We have learnt that real objectives are much more complicated than this; so are the eyepieces. Eyepieces can be divided into positive and negative: this is confusing! Most of the eyepieces that you are likely to meet will be negative eyepieces. What does this mean? This will now be explained.

The simplest negative eyepiece is the Huygenian eyepiece. It comprises two plano-convex lenses, each with the plane side uppermost. The rays from the objective are captured by the lower lens of the eyepiece and brought to a focus, forming the primary image *between* the upper and lower lenses of the eyepiece. This image acts, as we know, as the object for the upper lens of the eyepiece, which forms a virtual, secondary image of it. The upper lens of the eyepiece is acting just like the single lens we used to represent the eyepiece in our earlier diagrams. The lower lens of the eyepiece is "helping" in the formation, or rather in the placing, of the primary image.

The positive eyepiece, or Ramsden eyepiece, also comprises two plano-convex lenses, but in this case the plane surface of the upper lens is uppermost (as before), whilst the plane surface of the lower lens faces downwards. This combination acts as a simple magnifier. The primary image is formed, without any aid from the eyepiece, just below the lower lens of the eyepiece and acts as the object for the eyepiece combination. An enlarged virtual, secondary image is formed as expected. Both lenses of the positive eyepiece are acting in the same way as the single lens that we used to represent the eyepiece in our earlier diagrams.

So why are they called negative and positive? The answer is easy: the positive eyepiece acts as a simple magnifier, as does a biconvex, *i.e.*, positive, lens. If you use it to look at some print, the print will be enlarged. It does not matter whether you hold the eyepiece the right way up or upside down: it always acts as a magnifier.

The negative eyepiece cannot act as magnifier if used the right way up. The object that is normally magnified by the negative eyepiece (our old friend the primary image) lies between the upper and lower lenses of the eyepiece and we cannot put our piece of print there. But, if you turn the negative eyepiece upside down, it does act as a magnifier. You may have seen people use an eyepiece in this way to take a first, very low-power, look at a histological section before examining it with the microscope. Having a look at low magnification is fine; using an inverted eyepiece to do so is not. This is because it is easy to scratch the upper surface of the upper lens of the eyepiece on the coverslip. If you want to take a very low-power look at the section, use a pocket magnifier: a 10× magnifier is satisfactory.

The positive eyepiece is useful because a graticule for measuring structures seen with the microscope can be placed below the lower lens, just where the primary image occurs. The graticule will then be in focus when the section is in focus. A little shelf is provided and a threaded ring secures the graticule in place. Turn the eyepiece upside down, unscrew the locking ring, insert the graticule so that its sits on the shelf, replace the locking ring and turn the eyepiece the right way up again. A graticule can also be used with a negative eyepiece, but one of the lenses has to be removed to allow the graticule to be placed on a little shelf between the upper and lower lenses, just where the primary image is formed of course. Some eyepieces have a fine pointer cemented to this shelf: a helpful device. The pointer is in focus with the section and allows you to point out interesting features of the section to a colleague.

The eyepieces that have been described so far are satisfactory for use with low-NA objectives. For high-resolution work, compensating eyepieces are needed. Additional lenses are incorporated into the eyepiece to reduce any residual chromatic aberration left after correction of the objective. Both corrected negative and corrected positive eyepieces are available. Correction for curvature of field is also available: the "Periplan" eyepieces provided by Leitz are of this sort. It is always sensible to buy eyepieces to match your objective lenses.

People who usually wear spectacles often ask whether they should remove them when using the microscope. The microscope can compensate satisfactorily for abnormalities of refraction (short sight or long sight) but can do nothing for those suffering from astigmatism. Correction for astigmatism requires the use of cylindrical lenses. The author is short-sighted and also suffers from mild astigmatism: he removes his spectacles before using the microscope. For those who suffer from severe astigmatism and who therefore need to keep their spectacles in place, high eye-point eyepieces have been developed. These allow the user to keep the lenses of his or her spectacles away from the upper lens of the eyepiece, thus reducing the risk of damage to both.

Wide-angle eyepieces are also available. These are convenient when microscopes with a very wide field of view are used and when "plan" objectives are used.

3.5.1 The Exit Pupil of the Eyepiece

The rays that seem to come from the virtual, secondary image cross before they enter the eye. They form a small disc just in front of the eye; this is called the exit pupil of the eyepiece or the Ramsden disk (Ramsden of positive eyepiece fame). This should be about the same size as the pupil of the eye; if it is larger than the pupil of the eye, not all the secondary image will be seen. The diameter of the exit pupil is given by the formula provided by Hartley and shown in eqn 3.10:

$$D_R = 500 \times NA/M, \tag{3.10}$$

where D_R is the diameter of the Ramsden disk, NA is numerical aperture and M is total magnification. In practice this means that a total magnification of at least $170 \times NA$ is needed to keep the exit pupil of the eyepiece down to what Hartley described as a "reasonable size". Hartley suggested that an exit pupil of about 1.5 mm was satisfactory and produced the information described in Table 3.6.

Hartley made a further point: if the exit pupil is made too small, by over magnification, defects in the eye become visible as shadows on the image.

Binocular instruments are usually fitted with a focusable eyepiece. This allows differences in refracting power of the left and right eye to be adjusted for. The use of the focusable eyepiece is discussed below. It is also needed to correct for the change in tube length produced by narrowing or widening the gap between the eyepieces on a binocular microscope.

Table 3.6 Eyepiece magnification.

Objective magnification	NA	Useful magnification	Eyepiece magnification
5	0.1	33	7
10	0.25	84	8
40	0.65	215	5
60	0.85	280	5[a]
100	1.3	430	5

[a]In his table, Hartley puts 7 instead of 5 at this point; this may be a misprint in the original table.

We have now dealt with the essential components of the microscope: the objective, the condenser, the illumination system and the eyepiece. These are what form the image. A number of other, rather more mechanical components are also worthy of consideration.

3.6 THE LAMP HOUSING

It is very important to be able to centre the lamp to the optical axis of the microscope. This is checked by removing the lamp (housing and bulb) from the microscope and projecting the image of the filament, *via* the lamp condenser lens, onto the wall. Rotate the bulb mounting having first loosened the screw that holds it in place. The image of the filament should rotate around a single point. If it does not do so then it follows that it is not correctly centred and the centring screws must be adjusted to bring it back into line. Once it is aligned, tighten the locking screw and replace the lamp housing on the microscope. Even on modern microscopes with a limited range of adjustments available, it is essential to be able to align the lamp along the optical axis of the microscope. If this cannot be done, we must, again, trust that the manufacturer has done this for us. Whilst adjusting the lamp, never touch the bulb, especially if it is quartz halogen bulb. Grease from the fingers will be left on the bulb and will become "baked on" when the bulb gets hot.

3.7 THE STAGE

Most microscopes that are used for research are fitted with a mechanical stage that can be moved in two directions (from side to side, and to and from the operator) by means of rotating controls usually sited below the stage. The slide is generally held in place against a stop by a spring-loaded lever system. When placing the slide on the stage, ensure that it sits down properly onto the stage. If the spring-loaded lever has become bent upwards, it may lift the slide at one end. This is of course undesirable. The mechanical stage is calibrated so that the exact position of the stage at any given moment can be recorded. Note that the calibration system involves a vernier so that very precise recording of the position of the stage will be possible. This is a useful, but little used feature. A large stage is useful if large slides are to be examined. Piling slides that have been examined on the side of the stage is bad practice: they always fall of.

When examining sections during staining, it is convenient to use a microscope without a mechanical stage: this allows the section to be placed more rapidly on the stage and moved about with the fingers. Keeping a monocular microscope without a mechanical stage as the "staining microscope" is good practice. This also keeps staining solutions off the stage of the microscope used for research.

3.8   OPERATION

3.8.1   How the Sub-stage Condenser is Locked in Place

The condenser is often held in place by screws, by one screw, or is simply pushed into a tight retaining collar. When removing the condenser, always place a pad of cloth over the base of the microscope. It is very easy to drop the condenser whilst loosening the locking screw.

3.8.2   Centring the Condenser

This has been discussed already. Centre the image of the field diaphragm by focusing it in the plane of the section, partly closing it, adjusting the centring screws and checking that it expands evenly to fill the field of view.

3.8.3   Adjusting the Sub-stage Diaphragm and Recording its Optimal Position

This is done by means of a lever, which should project from the sub-stage condenser towards the operator. Close examination may reveal that the track of the lever is calibrated. It is possible to record the optimal position of the diaphragm by noting down the position of the lever when this is achieved. The author has not found this to be of much use. The author's Wild M20 EB instrument has a calibrated lever; the author's LeitzOrtholux microscope has not.

3.8.4   The Focusing Controls

Before using any microscope, make sure that you know what happens when you turn the coarse focus control towards you or away from you. On some instruments the focusing controls raise and lower the tube of the microscope; on others the tube is fixed and the stage is raised and lowered. On the author's Wild M20 EB, turning the focus control clockwise, *i.e.*, away from you, raises the tube and moves the objectives away from the slide. On the author's LeitzOrtholux, rotating the focus clockwise, *i.e.*, away from you, raises the stage and brings the objective closer to the section. Switching from one instrument to the other has caused confusion!

3.8.5   The Objective Carrier and Parfocality

Always arrange the objectives on the objective carrier in a logical order of increasing magnification. I find that turning the objective carrier clockwise to bring into play objectives of increasing power is intuitive. Perhaps this would not be the case if I were left-handed. On modern microscopes the objectives supplied by the manufacturer should be parfocal. This means that if the $10\times$ objective is in focus, you can turn in the $25\times$ and $40\times$ objectives and expect the image to remain in focus, or nearly so. You can also expect the $25\times$ objective to swing into place without it crashing into the surface of the coverslip. Such refinements should not be expected from old microscopes or from any microscope fitted with an "odds and ends" variety of objectives collected from other instruments. Check this before you begin to use the microscope.

3.8.6   Dealing with Dust and Dirt

Microscopists keep their microscopes in good condition; the same cannot be said for a significant proportion of histologists and pathologists. It hardly seems necessary to say that the microscope should be covered with a plastic cover when not in use. Dust and dirt really matters when it lies on a part of the optical path that is in, or is nearly in, focus with the section. The attentive reader will know that this means that dust is a problem when it lies in any of the conjugate planes that include the section. There are two places where this often occurs: dust may be on the slide or coverslip, or on the lenses of the eyepiece. Dirt on the front lens of the objective or on the lens of the sub-stage and field condensers causes less of a problem, though the quality of the image may be impaired. The presence of dirt on the lenses of the eyepiece may be confirmed by watching the dirt whilst rotating the eyepiece. If the dirt moves round with the eyepiece then it lies on the lenses of the eyepiece. Now unscrew the top lens of the eyepiece half a turn. Rotate only the top lens whilst watching the dirt: if it moves, it lies on the top lens; if it stays still, it lies on the lower lens. Having located the dirt you should remove it.

3.8.6.1   Removing Dust, Dirt and Greasy Finger Marks from the Surfaces of Lenses.   The best way to learn how to clean the optical components of a microscope is to watch the expert engineer from the manufacturer of the microscope do a routine service of the instrument. Three things are needed:

1. A rubber bulb and a narrow tube, which is used to puff air at a dusty surface. Notice that the expert does not breathe on the surface: there is no case for depositing water vapour on the lens.
2. A fine, dry, camel-hair brush is used to dislodge specks of dust that have resisted the blower. Barer noted a useful trick: hold the tip of the brush against a hot electric light bulb for a few moments. The brush will pick up enough electrostatic charge to attract dust from the surface of lenses.
3. Only if these measures fail, as they will if there is grease or greasy dirt on the lens, should lens tissue slightly moistened with xylene be used to clean the surface of the lens. Be careful: do not grind the dirt into the surface of the lens.

Cleaning the front lens of the objective is seldom needed. A greasy fingerprint may need to be removed with the lens tissue moistened, not soaked, with xylene. Immersion oil has to be removed from the objective and this should be done with dry lens tissue as soon as use of the lens is over. If immersion oil has been allowed to dry on the surface of the lens, it should be removed by repeated gentle wiping with the xylene moistened lens tissue. Never soak the front lens in xylene: the danger of weakening the cement that holds the components of the lens in place is real.

3.9 FIELD FINDERS

Few things are more annoying than noticing an interesting feature of a section and then not being able to find it again. Taking a reading from the "*x* and *y*" scales of the mechanical stage when the feature is being studied is an obvious solution. Note that a vernier scale is provided: this allows very accurate readings to be taken. Other methods are available. One involves the use of a field finder: a slide that has been engraved so as to divide it into small labelled squares. One well-known variety was and still is called the "England Finder". The section being studied is replaced by the field finder (without moving the mechanical stage of course) and the location indicated on the field finder noted down. To find the key area later, reverse the process. Another method is to stick a small arrow of paper onto the slide. Other methods, used in the past, are now seldom seen. One

such method was the use of a "marker objective", which was substituted for the objective in use and which carried a small rubber ring
that could be wetted with ink. The "marker objective" was then
lowered onto the coverslip and a small ink circle produced around
the point being studied. Lastly, "marker objectives" that carried a
diamond point could be substituted for the objective being used.
These were also lowered onto the coverslip and the diamond tip rotated so as to engrave a circle on the surface of the coverslip. I have
not seen such a marker for many years and have never used one!

3.10   USING THE MICROSCOPE

The reader that has struggled to here in this chapter may be relieved
to know that he or she is approaching the final section. Here, we
briefly review the steps in setting up the microscope for bright-field
observation. Of course, microscopes differ one from another and no
instructions that are both precise and universally applicable can be
provided. The golden rule when using a new microscope is: **read the
handbook before you start**. An older instrument, lacking a handbook
and with which you are unfamiliar, may present a few problems.
Remember to check how the focus controls work before you do anything else! We shall omit centring the lamp: see section 3.6 for this.
We shall also assume that the lamp is connected to the electricity
supply, is switched on and that the bulb works. If any of these assumptions are false then the remedies are obvious.

1. Place the section on the stage and centre it over the condenser
   so that some of the section lies under the objective. This is
   useful: in focusing, it is easier to find the section than to find
   the blank areas of the slide. Wiping the section with lens tissue
   to remove dust before you start is a sensible thing to do.
2. Ensure that the field and sub-stage diaphragms are wide open.
3. Focus the section using a 10× objective. There is no need to
   expect high image quality at this stage, all we need is the section in approximate focus.
4. Half close the field diaphragm and adjust the sub-stage condenser until a bright circle of light edged by the field diaphragm appears.
5. Centre the sub-stage condenser using the centring screws and
   open the field diaphragm until the circle of light fills the field of
   view.
6. Refocus the section.

7. Refocus the field diaphragm, closing it a little so that you can see the edge of the diaphragm, using the sub-stage condenser. Once perfectly focused, open it to fill the field of view.

8. If you can adjust the lamp condenser, focus the filament in the lower focal plane of the sub-stage condenser. Remember that adjusting the position of the lamp condenser lens does not alter the position of the field diaphragm. This step is often not possible when using microscopes with built-in illumination.

9. Close the sub-stage diaphragm slowly and watch the image like a hawk. Notice that it becomes a little darker and gains in contrast. Stop as soon as you think the quality of the image starts to deteriorate.

10. If the image is either too bright or too dark, adjust the lamp power control.

11. Remove one eyepiece and look down the tube. Look at the area of the back of the objective that is illuminated: the light should be filling about two thirds to three quarters of the available space. If the illuminated area looks less than this, open the sub-stage diaphragm a little, replace the eyepiece and take another look at the image. Only practice will allow you to decide when you have closed the sub-stage diaphragm just enough.

12. Now turn to the eyepieces. Ensure that the gap between the eyepieces is set so that on observing the section you can see a single circle of light and that this circle is as large as possible. Some workers have difficulty in fusing the images produced by the eyepieces. This is nearly always due to the eyepieces not being the correct distance apart. Set them widely apart and reduce the gap as you examine the section. Stop when you have a single large circular field of view. Then identify the focusable eyepiece. This is used to compensate, if necessary, for differences between the refractive powers of your left and right eyes. Close the eye that corresponds to the focusable eyepiece and focus the section using the other eye. Once it is in really sharp focus, close this eye and open the other. Now view the section through the focusable eyepiece. Adjust the eyepiece to bring the section into sharp focus. Open both eyes and view the section; a slight touch on the fine focus will now produce a first-class image. You will not need to adjust the inter-ocular spacing or the focusable eyepiece again.

13. If the high-power objective has a correction collar, now is the time to use it. Remember that the correction collar will be available only on high-power objectives and is designed to

correct for spherical aberration introduced by the thickness of the combination of the mountant and coverslip. Note that the correction collar is calibrated for the thickness of the coverslip (but not for the thickness of the mountant: this is assumed to be thin). Having produced the best image possible with the correction collar set at, say, 0.17 mm, adjust it slowly whilst observing the image. Look for improvements in the sharpness and contrast of the image. You will not notice a great improvement but, with practice, improvement of the image can be produced. If you are a real enthusiast, experiment with coverslips of different thicknesses and note the effects!

3.10.1 Setting up for Oil Immersion Microscopy

Assuming that you have the section in focus with the $10\times$ objective, decide whether you intend to oil the condenser to the slide. Most likely you will not do this because you will not have an oil immersion condenser to hand. If you are using a flip top condenser, make sure that the top lens is in place. Apply a generous drop of oil to the coverslip over the section using the glass rod from the immersion oil bottle. You will want to move the objective around in the oil: there is no place for tiny, pinhead drops of oil. Conversely, we do not want the oil dripping from the section. Do not plunge the glass rod up and down in the oil: this introduces air bubbles that are a nuisance if they lie in the oil above the section. Squeezy dropper bottles are useful for applying immersion oil. Now examine the section from the side: bend down and put your eye level with the section. Lower the objective to the oil or raise the oil to the objective. Do this by using the coarse focus control. As soon as the objective touches the oil, the oil layer will "light up". This is very obvious. Now examine the section through the eyepieces. You will not see much until you bring the objective even closer to the coverslip by use of the fine focus control. This is THE critical stage. Lack of attention can lead to you missing the section and impacting the objective on the coverslip. Go slowly. As soon as you see the section, focus it sharply. Now go back to the old routine: focus and adjust the field diaphragm, adjust the sub-stage diaphragm and adjust the light intensity if necessary. If you think you have missed the section, stop. Look from the side and rack the objective away from the section. Start again. You will not need to add more oil. If you rack back whilst looking down the tube of the microscope, you may miss the fact that the section may lift with the lens unless it is firmly held down by the mechanical stage.

If you are planning to use an immersion condenser, proceed as follows. Before you put the slide on the stage, lower the sub-stage condenser a little and put a good-sized drop of immersion oil on the top lens of the condenser. It will run over the top of the mount. Experience alone will tell you how large a drop to use. Now put the slide in place and raise the condenser. Watch from the side and note when the oil touches the under-surface of the slide. This is easy to see. If you raise the condenser too far, you may start to lift the slide. Make sure that the slide is nicely settled on the stage and proceed to immerse the objective. The adjustments of the condenser and diaphragms are as before.

Using what we might call the double immersion technique is the final test of your ability to use the microscope properly.

3.11 FLUORESCENCE MICROSCOPY

A substance that, when impinged upon by light of a given wavelength, gives off light of longer wavelength is described as fluorescent. Fluorescence microscopy takes advantage of this phenomenon and is, at least in outline, easy to understand. Light of short wavelength (towards the blue–violet end of the visible spectrum can be produced by a mercury vapour lamp or, more simply, by filtering out the longer wavelengths from the light produced by an ordinary tungsten filament lamp. The light is used to excite the fluorescent object: the filter, which filters the light going towards the object, is described as the exciter filter. Light leaving the object is also filtered so that only the longer wavelengths are used to form an image. The filter that does this is called the barrier filter.

Thus, the essential requirements for fluorescence microscopy are:

- A suitable lamp;
- An exciter filter;
- A barrier filter.

With these additions, a microscope set up for bright field microscopy can be used for fluorescence microscopy. The set up can be refined in two ways: the first is very straightforward, the latter is more complicated.

The first modification is to move from bright field to dark field microscopy. Dark field microscopy or, better, dark field illumination, employs a special condenser containing a circular black patch or "stop", which allows only very oblique rays of light to reach the object and only rays diffracted by the fine structure of the object to reach the objective lens (see Figure 3.7). These diffracted rays interfere with one another and form the image. The image appears bright, shining against a black background. The image is, in fact, the reverse of the image formed with the bright field set up, which is dark against a bright background. The great advantage of dark field illumination is the increased contrast—a very useful feature in fluorescence microscopy. Dark field illumination also allows objects that cannot be resolved to be seen. Very small objects, far beyond the limits of resolution (the usual rules limiting resolving power apply; see section 3.3.1), shine against the black background, as do tiny particles of dust in a beam of light striking across a darkened room. The structural details

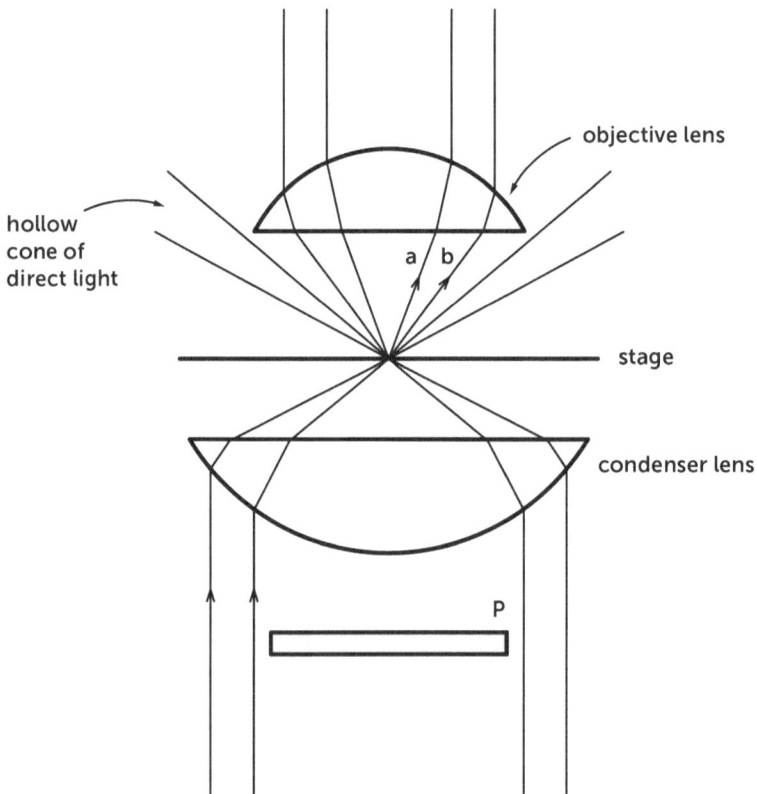

P: patch stop
a + b: diffracted light entering the objective

Figure 3.7　Dark ground illumination.

of these sub-resolution objects cannot be resolved, but the presence of the objects can be detected. This is not important for us; the increased contrast provided by dark field illumination is the key feature for fluorescence microscopy.

The second modification is more dramatic. The light path is changed so that light of suitable wavelength passes down through the objective to the specimen, and light of longer wavelength (produced by the fluorescent object) passes up through the objective and is used for image formation. This set up is described as epi-illumination and was invented by the distinguished microscopist, J S Ploem. Figure 3.8 shows the set up.

The key to epi-illumination is the dichroic mirror. This rather special mirror has unusual characteristics: it reflects light of certain wavelengths whilst allowing light of other wavelengths to pass straight through it. A very special mirror indeed! Note from Figure 3.8 that the dichroic mirror is placed above the objective lens. The exciting light passes down from the dichroic mirror through the objective, which acts, rather nicely, as its own condenser. The longer wavelength light coming up from the object is passed through the dichroic mirror to the eyepieces: this is the light that forms the image. This set up allows the object to be illuminated (excited) with light of appropriate wavelength and for the image to be formed only by the

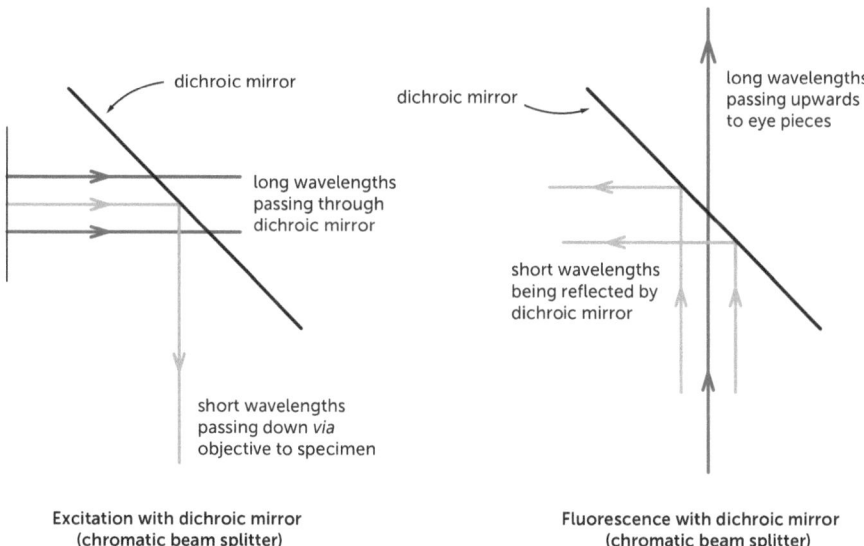

Excitation with dichroic mirror
(chromatic beam splitter)

Fluorescence with dichroic mirror
(chromatic beam splitter)

Figure 3.8   **The dichroic mirror or chromatic beam splitter.**

longer wavelength light produced by the fluorescent object. One could hardly ask for more! The addition of a barrier filter above the dichroic mirror completes the process.

It will be obvious that exciter and barrier filters and the dichroic mirror, if epi-illumination is used, must be carefully chosen to match the needs, so to speak, of the object. An appropriate wavelength is needed for excitation; only appropriate wavelengths must be used for image formation. Choice of filters and dichroic mirrors will depend on the fluorescent material being examined. If the material produced a green fluorescence then, for example, a different barrier filter will be needed as compared with an object producing a red fluorescence. Catalogues should be consulted for details. Good accounts of the details of fluorescence microscopy are provided by the works listed in the Bibliography.

CHAPTER 4

How to Examine Histological Sections

Learning how to examine histological preparations ("sections" or "slides") is a key step in learning about histology. Rather oddly, it is often ignored or touched upon by teachers only very lightly. One simply looks! Nothing could be further from the truth: learning how to look, how to recognise and how to identify changes from normality are absolutely essential. The approach outlined here makes few assumptions about the reader's experience. We begin at the beginning and develop a strategy for examination that can be applied to all histological preparations. Learning how to recognise tissues is always a problem for beginners who ask how the experienced histologists can distinguish, at a glance, one tissue or organ from another. The answer is, to a large extent, practice, but practice needs to be guided and we have tried to show how that can be done. Some reference material dealing with types of epithelia has been included to help the beginner.

4.1 INTRODUCTION

The examination of histological sections is the last stage of a long process: it should never be seen as the only important stage of the process! The experimental animal (I am assuming that this book will be read, if it is read, by those doing experimental work rather than by clinical pathologists) will have been closely observed whilst alive, a careful autopsy will have been performed and the findings recorded; blocks of tissue will have been taken and examined with the naked eye: all this before the sections come to microscopy.

Histological Techniques: An Introduction for Beginners in Toxicology
By Robert Maynard, Noel Downes and Brenda Finney
© Maynard, Downes and Finney 2020
Published by the Royal Society of Chemistry, www.rsc.org

Nearly all older works advise that sections (using this as short hand for histological sections) should be examined with the naked eye or with a hand lens before being examined with the microscope. This excellent advice is widely ignored. This is an error because useful information can be gained by very low-power examination: large areas of consolidation of the lung can be easily recognised, the distribution of lesions can often be appreciated and the tissue being examined can often be identified. It would be sensible to see the examination of sections as a progression from low to high power and then, if necessary and using different material, on to electron microscopy with the superb resolution that that technique offers. Rushing to high power is a cardinal error: much more is likely to be learnt at low–medium power. As one distinguished American pathologist said, "What we need are higher powered pathologists using lower powered microscopes!"

4.2 SCANNING

The first and most important step in examining any section is to examine ALL the section at low power. Only in this way can you be sure that salient features are not missed. The section should be examined in an orderly way. Start at a corner or edge and scan across horizontally using the mechanical stage. Then move vertically and scan back across, *i.e.*, horizontally. Continue until you have seen all the section. You will see why a very low-power objective, which allows all the section to be seen without moving the mechanical stage, is useful. Such objectives are available, for example, 2.5× and 5×. Please refer to Chapter 3 for advice on how to use these objectives: you may need to remove or replace the condenser on your microscope. There is much to be said for using a high-quality, low-power, dissecting microscope for this initial scanning, but few workers seem to do this.

Whilst scanning, you will be observing and confirming that the tissue is what the label on the slide says it is. This may seem a trivial point but errors do occur. You should also be bearing in mind how the tissue was processed. Chapters 5 and 6 provide information on artefacts produced by, for example, different fixatives and paraffin wax embedding. The large spaces often seen around blood vessels are shrinkage artefacts produced by paraffin embedding: they are not seen in celloidin sections. You should also be looking for artefacts introduced during sectioning, during attaching of sections to slides and during mounting. Cracks in the section, score marks, chattering marks, folds and bubbles are all possible and should be noted. Look, too, for the tell-tale budding spores that are produced by yeasts that may have contaminated the mounting medium.

4.3 EXAMINING BLIND: "BLINDING THE PATHOLOGIST"

Most pathologists like to know, before they look, what they are looking at. Of course, this is sensible and sections are carefully labelled to allow their provenance to be traced. This is only good practice. But there are occasions, for example, when minor changes between, say, tissues from animals exposed to a low dose of some chemical and control animals are being sought, when examining blind is helpful. The objective is to remove bias. It is possible that if one knows that a section is from a control animal that one will not examine it as carefully as one from an exposee. Thus, minor changes in the controls can be missed and similar changes in the sections from the exposees can be regarded as more important than they actually are.

4.4 RECOGNITION OF TISSUES: DIAGNOSIS OF SECTIONS

Tests of one's ability to recognise and describe sections of tissues used to be a standard part of examinations in histology and still are in some courses on histopathology. For those students who would never again be asked to identify a section, these tests seemed pointless; for histopathologists, the ability being tested is really rather important. Recognising the tissue of origin of a metastasis, for example, depends on one's knowledge of the appearance of normal tissues. Being able to tell "at a glance" whether a section reveals normal or abnormal tissue depends on a well-developed grasp of the normal.

Recognition is usually much easier at low power than at high power. Take, for example, a section of kidney as shown in Figure 4.1. A high-power view of a few tubules might be recognisable by a histologist but a low-power view that shows the cortex and the medulla, the glomeruli of the cortex and the bundles of tubules dipping down into the medulla will be instantly recognisable by almost anybody. Scanning at low power is strongly recommended if asked to identify a section. We shall approach the problem in three steps in the following sections.

4.4.1 What is the Tissue?

Two methods of identification are available: the "instant visual recognition method" and the "analysis of components" method. These methods are by no means entirely separate: recognising and identifying components relies on visual memory.

4.4.1.1 Instant Visual Recognition. This is probably how all experienced histologists identify sections of tissues. Histologists are likely to recognise, immediately, a section of liver or kidney or lung. The overall pattern is embedded in the memory and is summoned up on

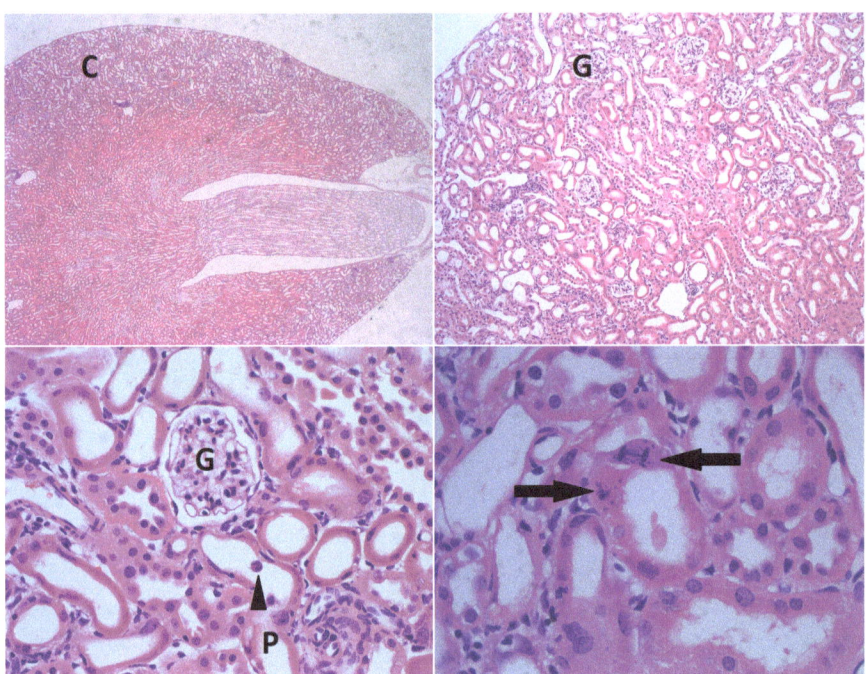

Figure 4.1 **Different information can be obtained from different magnifications.** At the lowest power (top left) it is clear you are looking at kidney and there looks to be some pathology in the outer cortex (C). At a higher power (top right) you get the impression that the lesion is confined to the proximal nephron as the collecting ducts and pars recta look to be unscathed. Up another power level (bottom left) and you can see the glomerulus (G) is also healthy enough and, in fact, the proximal convoluted tubule (P) is the primary target. There is some acute necrosis and an element of recovery; note in the annotated proximal tubule the wall thickness is not uniform and there is a necrotic epithelial cell floating down the lumen of the proximal tubule (arrowhead). If you look for long enough at a high power (bottom right), you may be rewarded with something exotic like this metaphase and telophase occurring side by side (arrows).

viewing the section. Recognition, like recognition of one's house or motor car or partner, is practically instantaneous. This is an admirable and enviable skill. Those with good visual memories acquire this skill rapidly; those with poor visual memories struggle to do so. Practice, as ever, is the key to acquiring the skill. But herein lies a difficulty: tests of one's visual memory may occur before enough practice has been possible. Under the pressure of such tests the temptation to guess can be strong and errors are inevitable. In what many regard as the "bad old days" of teaching of anatomy and histology, when a much more detailed knowledge was required than is now the case, students' propensity to guess provided their tutors with a steady source of entertainment. Sections of oesophagus were described as sections of vagina; the prostate was identified as foetal lung or lactating mammary gland and so on. The panic-stricken student, having decided on the diagnosis, would "find" features to support it and plunge deeper and deeper into the mire. But even if guessing is excluded there are several explanations for failure to identify a section by the "instant recognition method".

Some tissues look rather alike and, unless components are examined, errors are very likely. For example, distinguishing the parotid gland from the submandibular gland is not difficult for the expert. But it cannot be denied that both are exocrine glands, indeed both are salivary glands, both have acini and ducts, and both might be mistaken for each other by any beginner hoping to decide which is which "at a glance".

Instant visual recognition depends on certain features being present and appreciated unconsciously. Thus, someone who habitually wears a moustache might be difficult to recognise after he has shaved it off, though his appearance has changed only in this one detail. It would certainly be easy enough to mistake a section of pancreas for one of the parotid gland if the former lacked islets of Langerhans. The importance of scanning at low power is obvious, looking for the islets is very important.

Damage to the normal appearance of the tissue can easily produce confusion. The consolidated lung does not, at first glance, look much like normal lung. Only on detailed examination will the key feature of recognition appear. This is, of course, the presence of cartilage in the walls of the bronchi. These bronchi might, at a glance, have been thought to be the ducts of some gland.

Unfamiliar planes of sectioning can greatly alter the appearance of sections of tissue. Anybody can recognise a section of skin in which the plane of the section is vertical to the surface: the layers give away the diagnosis. But a horizontal section that revealed pale islands, comprising cross sections of dermal papillae, amongst a darker cellular background formed by one of the deeper layers of the epidermis might confuse almost anybody. It will be clear that sticking to a routine for sectioning tissue in given planes is important. It will also be clear that thinking in three dimensions is important: there is more to the architecture of a tissue than is revealed in a single section!

4.4.1.2 Analysis of Components Method. As pointed out above, this method does not do away with the need for visual memory but it does allow the diagnosis of the tissue to be reached by the application of logical reasoning. A complete system has not been developed: it would be immensely detailed and complicated. Some tissues can be recognised at a glance and the observer is unlikely to be confused between these and other tissues. The approach outlined here is a guide, but not more than that. The guide proceeds by asking and answering questions. Much of the supportive material is in note form.

4.4.1.3 Tissue Recognition. **Tissues to be recognised at a glance:** note that here we are recognising tissues rather than organs.

- Bone;
- Cartilage: hyaline, white fibro-cartilage, elastic cartilage;
- Tendon;
- Yellow elastic ligament;
- Nervous system: neural tissue, as in spinal cord and brain.

Organs to be recognised at a glance: here the shape or architecture of the organ gives away the diagnosis.

- Eye: cut as a whole so as to include the lens, iris and cornea;
- The semi-circular canals and cochlea of the inner ear;
- A cross section of the nose of a rodent: the complexly scrolled conchae give away the diagnosis;
- A section of the rodent heart: the cavities (ventricles and perhaps atria), the inter-ventricular septum, the valves and the cardiac muscle in the rather thick walls cannot be confused with any other organ.

Organs and structures that may require more effort: begin by deciding whether the section is of a solid tissue, perhaps liver or spleen or kidney or testis or a gland of some sort, or of something with a lumen, for example, gut, uterus, fallopian (uterine) tube, vas deferens or ureter. Organs with a lumen have a specialised luminal surface: the epithelium is characteristic of the organ. For example, the trachea has a pseudostratified ciliated columnar epithelium, whereas the bladder has a multi-layered transitional epithelium. If the epithelium is ciliated, there are not many things that the organ could be: trachea or bronchus, fallopian tube, the uterus, part of the epididymis or the vas deferens (ductus deferens). The enthusiast would wish to point out that the ependymal cells of the central nervous system are also ciliated, but this is unlikely to be the way in which you decide that the section is a section of brain. If cartilage can be found in the wall of what you will now be thinking of as a tube then it can only come from the lung: trachea or bronchus.

Solid structures: when faced with a solid structure, something that is all cells and without a specialised surface to ponder about, you will know, at least, what it is not! A rectangular section of tissue with no luminal surface is not a section of gut! Of course, it would be possible to cut a section of gut, cutting horizontally along the wall, in such a way that no lumen could be seen. We shall ignore such eccentricities. Is the tissue a gland of some sort? Glands secrete: they have secretory cells. Exocrine glands have ducts: the ducts lead to a surface where the secretion is discharged. The salivary glands or the lacrimal gland provide examples. Endocrine glands secrete direct into the blood stream. Exocrine glands come in a variety of shapes and forms but all have a secretory part and a duct system. The secretory part may be arranged as straight or branched tubules or as collections of cells grouped around the end of the duct system. The glands of the stomach are tubular glands; they are branched tubular glands and open into the gastric pits, which are themselves narrow tubules that run down from the surface of the mucosa. In those glands in which the secretory tissue is grouped around the end, the distal part, or the duct system, the secretory cells may be arranged as acini or as alveoli. There is not much difference between the two but acini show essentially no lumen and alveoli always show a small lumen. One could search all day and not find a lumen in the acini of the parotid gland; the alveoli of the lacrimal gland are immediately obvious.

The liver and kidney could both be described as glands: the liver secretes bile and the kidney excretes urine. Each has a secretory portion: the hepatocytes of the liver and the glomeruli of the kidney. But nobody would really call the kidney a gland in the sense that the pancreas or thyroid are glands. Similarly, the testis contains a multitude of seminiferous tubules that produce spermatozoa: one would not describe this as a secretion, though one could. The point being made is that organs that produce things for release to a surface have ducts. The kidney contains not much except ducts of one sort or another; the liver has ducts, though these are not so obvious; the testis has many ducts that are collected together, so to speak, at the epididymis and the secretion is channelled to the vas deferens (ductus deferens, if you are up to date on your terminology, though vasectomies are commoner than "ductectomies").

Other solid organs are full of lymphocytes. These are the organs of the lympho-myeloid complex, and include the thymus, the lymph nodes and the spleen. The tonsil also falls into this category, though it is rather special in that it has a surface that lines part of the pharynx. Nobody could fail to tell a lymph node from a lung, well, at least from a normal lung, but distinguishing a section of a lymph node from a section of thymus might be thought to be trickier. In fact, it is not at all difficult: see below.

4.4.1.4 Recognition by Examination of Surfaces. By this we mean recognition by being able to identify the epithelium that lines the working surface of the organ or structure. Working surface? What does that mean? Recall that the gut has two surfaces: a luminal surface, where the work of the organ (secreting, absorbing) is done and an outer surface that simple surrounds the organ. The outer surface of the gut comprises a simple mesothelium (an epithelium derived from mesodermal tissue) and is a part of the visceral peritoneum. This surface is not without its own interest, but we are focusing on the lumen—the busy hard-working surface of the gut. Examination of the luminal surface of any organ (it must be a hollow organ to have a lumen so we can exclude solid organs such as the liver or kidney) will reveal an epithelium; identification of the organ begins with identification of that epithelium. Knowing the different varieties of epithelia found in the body is a key part of knowing about histology. If you don't know the headline terms, or any of the other terms, please consult a textbook or atlas of histology (see Bibliography).

Although there are a few specialised types of epithelium, the majority of epithelial tissues can be categorised in terms of the shapes of the cells and the manner in which the cells are layered. The morphologies that we can recognise are squamous, cuboidal, columnar and transitional. There are three principal morphologies associated with epithelial cells. Squamous epithelium has cells that are wider than they are tall. Cuboidal epithelium has cells whose height and width are approximately the same. Columnar epithelium has cells taller than they are wide.

Simple epithelium consists of a single layer of cells. Given the delicate nature of this type of surface, it is unsuitable for areas that are prone to undergo wear and tear. Since simple epithelium is one cell thick, every cell is in direct contact with the underlying basement membrane. It is generally found where absorption and filtration occur as the thinness of the epithelial barrier facilitates these processes. The various types of epithelium that are most commonly encountered are described below.

Simple squamous epithelium

Simple squamous epithelium consists of a single layer of flat, scale-like cells, adapted for diffusion and osmosis. It is widely seen in the endothelial lining in the heart, blood vessels, lymph vessels and as the walls of capillaries. The other most commonly encountered areas for this epithelium are the alveoli of lungs (shown in Figure 4.2) and parts of the loop of Henle in the kidneys.

Stratified squamous epithelium

Stratified epithelium differs from simple epithelium in that it is multi-layered and is also shown in Figure 4.2. When cells are stacked in several layers, the epithelial surface has a far higher resistance to wear and tear. As such, it is found where surfaces have to withstand mechanical or chemical insult, where the surface layers can be abraded and lost without exposing sub-epithelial layers or delicate underlying tissues. The top layers are quite flattened and may be covered by

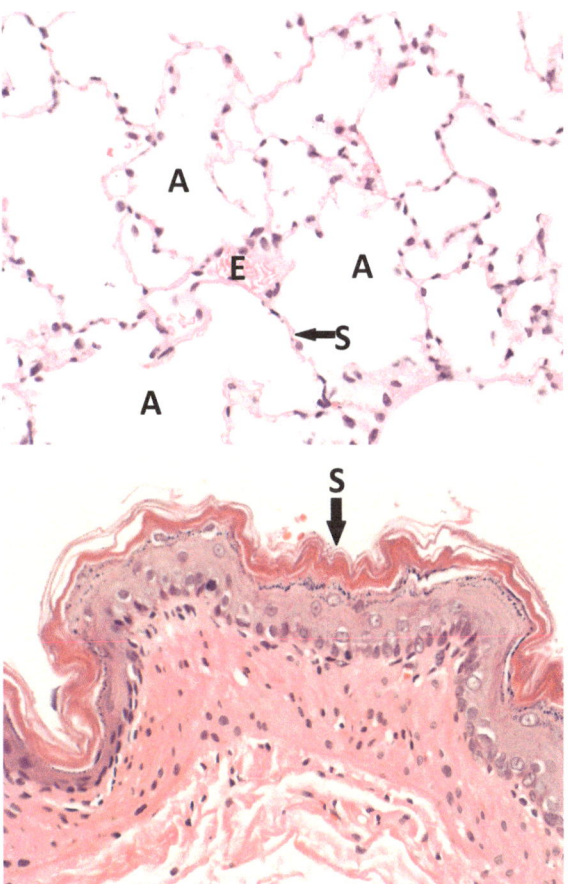

Figure 4.2 Squamous epithelium. Simple squamous epithelium in concert (top panel). The simple squamous epithelium (S with arrow) of the lung alveoli (A) is in close approximation with an endothelium (E) lined capillary. As you can see, there is very little connective tissue in between these cells to interfere with gaseous exchange. Stratified squamous epithelium (S with arrow, bottom panel) in the stomach of a rat.

keratin in areas of very high abrasion. The lower layers are a little more rounded and the basal layer is generally very active, continually replacing the layers above. As the older cells are pushed to the surface and further away from blood supply, the cells get flattened, dehydrated and harder. Once on the surface, they eventually get rubbed off.

In keratinized epithelia the most apical layers (exterior) of cells are dead and lose their nucleus and cytoplasm, and instead contain a tough, resistant protein called keratin. Keratin is a fibrous structural protein and stains quite a bright orange colour in H&E sections. The keratinized version is very friction resistant, waterproof and resists bacterial infection, so is ideal for mammalian skin. It does vary quite considerably between various sites and between species. In man it is thinnest on the scrotum and thickest on the soles of the feet. For comparative purposes, the pig is probably closest to man in structure.

The non-keratinized stratified squamous epithelium is found on moist surfaces exposed to wear and tear, such as the mouth and tongue, the lining of the oesophagus (although this is keratinized in rodents) and the lining of the vagina.

Simple cuboidal epithelium

Simple cuboidal epithelium comprises closely fitted polygon-like cells with a centrally located nucleus. It is most often seen lining small ducts. Other sites where cuboidal epithelium may be seen include parts of the nephron in the kidney, the choroid plexus in the brain and in lining the follicles of the thyroid gland, as shown in Figure 4.3. This type of epithelium is also found in areas where secretion and absorption are prime functions.

Simple columnar epithelium

Simple columnar epithelial cells appear rectangular in cross section and have nuclei at the base of cells, as shown in the bottom panel of Figure 4.3. It is uni-layered and is typically seen lining the gastrointestinal tract, gall bladder and the excretory ducts of many glands, but they frequently have modifications to enhance their function

The most commonly seen modification is ciliation. In routine preparations cilia can be difficult to see, although in perfused fixed specimens it is easier. Cilia acting in unison can also waft substances along the surface of the cell. These are seen in the walls of trachea and bronchi, the uterus and fallopian tubes, as well as the ventricles in the brain.

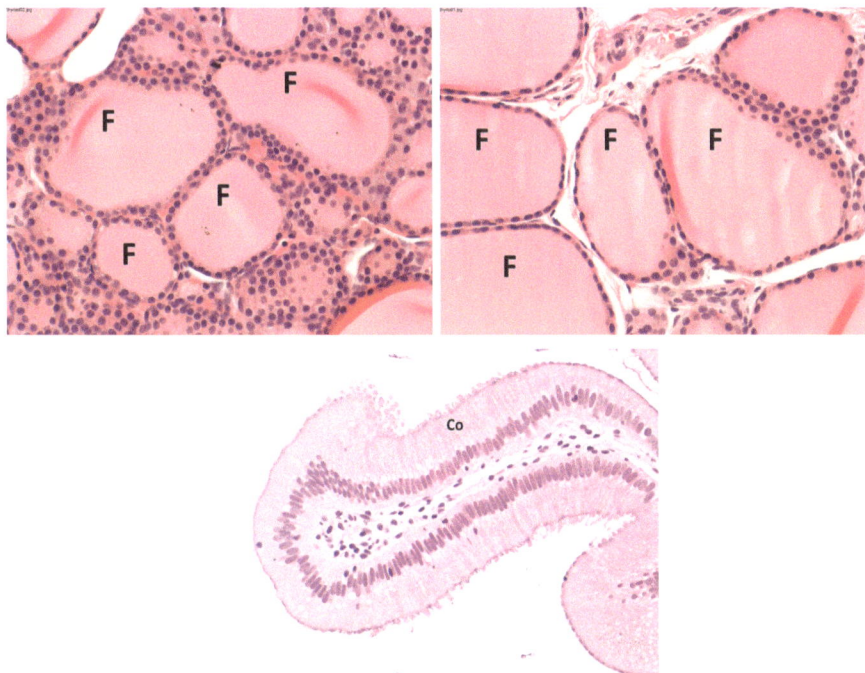

Figure 4.3 **Cuboidal and columnar epithelium.** Cuboidal epithelium (top panel) lining the follicles of the thyroid gland (F). Although cuboidal epithelium does not have the stretch that transitional epithelium has, it does have some flexibility. In the top right panel of this thyroid gland you can see that the epithelium lining the distended follicles is quite flattened and looks more squamous than cuboidal. Columnar epithelium in the gall bladder (Co, bottom panel). Notice that the nuclei are all at about the same position in the cells, although at the tip of the fold it does have a slightly stratified look to it. This is purely artifact and the result of an oblique angle of cut.

In the gut, the surface of the columnar cells has microvilli that serve to increase surface area, and thus aid in absorption. Another common modification is goblet cells that secrete mucus, which serves as a lubricant or protective layer.

Pseudostratified epithelium is really a variation of the simple epithelium as it consists of just a single layer of cells, but looks as though it has multiple layers because the nuclei lie at differing depths within the cells. All the cells are attached to a basement membrane, but not all cells reach the free surface. As it rarely occurs as squamous or cuboidal epithelia, it is usually considered synonymous with the term pseudostratified columnar epithelium and is exemplified by the section in the top panel of Figure 4.4. This epithelium often has the same

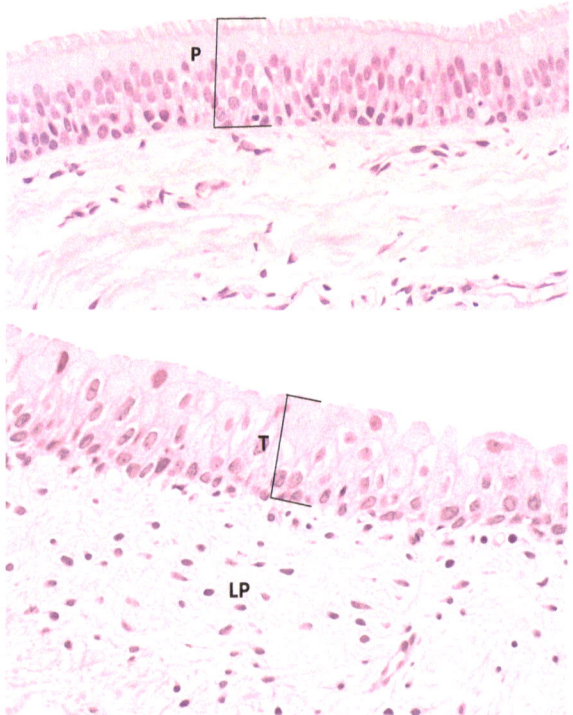

Figure 4.4 **Pseudostratified and transitional epithelium.** Pseudostratified epithelium (P with bracket, top panel) from the trachea. Looking at this it is difficult to believe that all these cells are resting on the basement membrane. Transitional epithelium (T with bracket, bottom panel) in the urinary bladder, which sits on top of a loose lamina propria (LP).

modifications of the simple columnar epithelium and is often ciliated or mucous-secreting. This type of epithelium lines large areas of the respiratory tract, where the beating cilia direct the mucous secreted locally by the goblet cells up and out, rather than allowing it to sink down to the alveoli.

Transitional epithelium

Transitional epithelium contains cells of various shapes and is found on structures that can change shapes during certain body functions. It is sometimes called the urothelium since it is exclusively found in the bladder, ureters and urethra. These surfaces have tremendous extensibility and elasticity. The appearance can vary depending upon the degree of distension in the areas in question. It may appear to be stratified cuboidal when the tissue is not stretched or stratified squamous when the organ is distended. An unstretched sample from the bladder is shown in the bottom panel of Figure 4.4.

There are some other quite rare variants, such as stratified cuboidal epithelium and stratified columnar epithelium, but these are so uncommon as to be not worth worrying about (Table 4.1).

Table 4.1 Varieties of epithelia.

Structure	Epithelium	Structure of the wall	Additional features
Oesophagus	Stratified squamous, keratinized in rodents but not in man	A muscularis mucosa is present. The muscular layer proper is arranged in two distinct layers: the inner is circular, the outer is longitudinal. In the upper oesophagus the muscle may be skeletal rather than smooth	Small glands will be seen in the submucosa
Ureter	Transitional	Muscle less clearly arranged in layers than in oesophagus. Three layers: inner longitudinal, middle circular and outer longitudinal	
Urethra	Columnar: stratified or pseudostratified. No cilia or stereocilia	Muscle not arranged in clear layers	The lamina propria contains a prominent plexus of small veins
Ductus (or vas) deferens	Columnar, pseudostratified with sterocilia	Very thick muscle layers: inner longitudinal, middle circular and outer longitudinal	The ductus deferens is often sectioned as part of the spermatic cord: look for the other contents of the cord
Uterine tube	Columnar, simple with cilia	No obvious layering of the muscle in the wall, but it tends to be arranged circularly	The mucosa is folded; the folds are thin.

It will be seen that the epithelium is the key to the diagnosis.

Glands often present a problem. Species differences need to be considered when looking at salivary glands. Before beginning research that involves a study of the effects of a chemical on the salivary glands it would be sensible to make a careful study of the histology of these structures. Building up a collection of normal, control tissue and sections is essential.

Glands can be divided into those with acini and those with alveoli (see above). In Table 4.2 we shall consider the salivary glands, the lacrimal gland and the Harderian gland. The latter does not occur in man, but is found in the rat.

It will be appreciated that special stains would be of great help in distinguishing the glands listed in Table 4.2. The periodic acid–Schiff (PAS) technique for mucin and the Oil Red O technique for lipids would identify the mucus-secreting salivary glands and the Harderian gland, respectively.

Table 4.2 The histology of glands.

Gland	Secretory units	Ducts	Other features
Parotid	Serous acini only. The cytoplasm of the acinar cells is deeper stained than that of the other salivary glands. No mucous acini at all	Many profiles of ducts are visible. Many striated ducts are seen. The intercalated ducts that are easily seen in man seem less common in the rat. Ducts look pale in comparison with the acini	In man the facial nerve runs into and divides in the gland. Sections of nerve may be visible. Fat and lymph node may also be seen
Submandibular	In man this is a mixed gland with areas of serous acini and areas of mucous acini. The mucous acini are often surmounted by serous demilunes. These used to be called the serous demilunes of von Ebner or of Heidenhain or the serous crescents of Giannuzzi	Fewer profiles of ducts than in the parotid. Striated ducts present. The brightly eosinophilic secretory ducts of the male rat are diagnostic. Fewer are seen in females	

Table 4.2 (*Continued*).

Gland	Secretory units	Ducts	Other features
Sublingual	In man this is a mucous gland, serous demilunes can be seen and, now and again, a serous acinus. The cells of the mucous acini are pale and vacuolated. The cell margins stand out here and there against the vacuolated interior of the cells. Darkly stained, small nuclei squashed against the base of the cell with their long axes horizontal. Though the secretory unit is an acinus, very small (about the size of a nucleus), nicely rounded spaces can often be seen at the centre of the acini: like a mini-alveolus	Ducts more darkly stained than the acini. Intercalated ducts are rare in man. Striated ducts are also rare	
Lacrimal. The rat has two lacrimal glands: one infraorbital, close to the Harderian gland, and one extra-orbital, close to the parotid. Man has only one: high up in the upper lateral corner of the orbit. (An accessory gland also occurs in man but we shall ignore that.)	In man the lacrimal gland is an alveolar gland: distinct, if small, alveoli can be seen in the secretory units. In the rat the gland is described as acinar. The cells of the acini stain less densely than those of the parotid. No mucous acini	No brightly staining secretory ducts, as in the submandibular gland. Fewer duct profiles visible than in the parotid	

Table 4.2   (*Continued*).

Gland	Secretory units	Ducts	Other features
Harderian gland. Not found in man but present in the rat. Lies within the orbit, wrapped around the optic nerve	Alveoli rather than acini present. The epithelium is columnar with clearly defined nuclei towards the bases of the cells. Some alveoli contain dark brown staining material containing porphyrin: these are not seen in any of the other glands described in this table. The apical surfaces of the cells are domed. The secretion contains lipid; the secretion is merocrine and not apocrine as in the mammary gland or holocrine as in the preputial gland	No special features: cuboidal–columnar epithelium	A section of the optic nerve or of extraoccular muscle may appear. Once seen, it would be difficult to mistake this gland for a salivary gland
Exocrine pancreas. This is very like a salivary gland in appearance and is included for completeness	Serous acini, no mucous acini at all	Fewer duct profiles than in the parotid. No striated ducts but intercalated ducts are present	The key diagnostic feature is, of course, the presence of islets of Langerhans in the pancreas. Centro-acinar cells are seen in the pancreas but not in any of the salivary glands

A selection of alveolated glands that sometimes cause difficulties in recognition

Reasonable care will lead to blocks of prostate not being mixed up with blocks of thyroid. However, at first glance, their histological appearances are not dissimilar and they, and a few other glands, have been included in Table 4.3. The foetal lung **is** a gland: it secretes fluid.

Table 4.3

Gland	Secretory portion	Duct system	Other features
Prostate	Alveoli lined with columnar epithelium. In the rat the alveoli appear larger than in man (proportionally) and look more like thyroid follicles. Projections from the walls of the alveoli push out into the alveolar spaces. This feature is not seen in other alveolated glands	Hardly any profiles of ducts seen	In man the critical diagnostic feature is the presence of smooth muscle in the septae between the secretory units. The secretion may condense to form prostatic concretions, which are dense and may be lamellated in appearance
Lactating mammary gland	Secretory units vary considerably in size: this is striking. Milk may be seen in the lumens; fatty globules seen in the cells. A lipid stain would show these well	Ducts seen in the inter-lobular septae	Very clearly divided into lobules by fibrous tissue. Unlike the prostate, no smooth muscle in the septae
Thyroid	Follicles lined by cuboidal epithelium and contain colloid. The epithelium varies in height with the activity of the follicle: high in the resting follicle, low—almost squamous—in the very active follicle. The colloid stains bright pink with H and E and a mixture of red and blue with the trichrome stains (assuming that a blue stain, rather than a green stain, was included!) This is an artefact	NO DUCTS, this is an endocrine gland	Sections of thyroid often contain parathyroid gland, as well as thyroid. The parathyroid is quite different from the thyroid: no follicles, with epitheloid cells arranged in clumps between delicate strands of connective tissue. No ducts of course! An occasional droplet of colloid can be seen: why this is so I have no idea
Foetal lung	Branching tubules rather than alveoli. Lined by cuboidal epithelium. No proper alveoli, in	The "duct system" is represented by the conducting airways. The larger airways contain	

Table 4.3   (*Continued*).

Gland	Secretory portion	Duct system	Other features
	the respiratory sense, at this stage	hyaline cartilage in their walls: this is THE diagnostic feature. The epithelium of the larger airways is pseudostratified ciliated columnar	
Preputial gland. Found in the rat but not in man	A modified sebaceous gland. Holocrine secretion. Lining of the secretory portion: stratified epithelium. Secretory cells are foamy, containing eosinophilic granules	Large central duct: stratified, sometimes keratinized epithelium	Quite different from a gland with acini or alveoli lined with cuboidal or columnar epithelium. Not to be confused with Cowper's glands (human male) or the equivalent Bartholin's glands (human female), both of which are mucus-secreting

Identification of different parts of the gut

Table 4.4 summarises the identification of gut tissue.

Staining for individual components of the tissues of the gut clearly has a lot to offer. Consider:

- Staining for mucin (PAS or Alcian Blue);
- Staining for granules in Paneth cells;
- Silver staining for argentaffin and argyrophilic cells;
- Staining for lipids in lacteals;
- Staining for connective tissue and smooth muscle.

See Chapters 7 and 8 for more detail.

Lymphoid tissue

The lymph node, spleen and thymus must be carefully distinguished. Each contains aggregations of lymphocytes and, at first glance, look rather like one another. In actual fact distinguishing these structures is easy.

Lymph node. Lymphoid tissue occurs in nodules; if active, these have a pale centre and a darker staining surround. The lymph node is

Table 4.4 Regions of the gut.

Region	Epithelium	Glands	Other features
Oesophagus	Stratified squamous, keratinised in rodents but not in man	Oesophageal glands proper: mucus-secreting, in the submucosa. Mucosal glands that look like gastric glands at the upper and lower ends of the oesophagus: so called cardiac glands	Muscularis mucosa present. Two layers of muscle, deep to submucosa: inner circular and outer longitudinal. The lamina propria pushes up as papillae into the under surface of the epithelium (rather like skin)
Fore-stomach	Stratified squamous keratinized. An extension of the oesophageal epithelium into the rodent stomach		
Stomach (proper)	Fundus: columnar epithelium. Surface is mucus-secreting but no goblet cells appear	Deep gastric pits. Tubular glands open into the pits. Parietal cells produce acid: markedly more eosinophilic than other cells. Chief cells producing enzymes.	Muscularis mucosa sometimes divisible into an inner circular and an outer longitudinal layer: not easy to see! Muscularis externa (main muscular layer): inner oblique, middle circular and outer longitudinal
	Pylorus	Deep pits, pyloric glands open into pots. No parietal cells, no chief cells. Mainly mucus-producing	
Duodenum	Villi with crypts of Lieberkuhn in between them. Goblet cells present	Brunner's glands in the submucosa. Apart from the oesophageal glands proper, these are the only submucosal glands of the gut	From here on, the muscularis externa is clearly divisible into inner circular and outer longitudinal layers
Small intestine	Villi, crypts and goblet cells are all present	Only those of the mucosa	Muscularis mucosa sends strands of muscle up into the villi. Aggregations of lymphoid tissue in the ileum (Peyer's patches), which extend into the submucosa

Table 4.4 (*Continued*).

Region	Epithelium	Glands	Other features
Large intestine	No villi. Crypts of Lieberkuhn present many goblet cells. Surface looks flat in comparison with the small intestine	Only those of the mucosa	In man the inner circular layer of the muscularis externa is complete but the outer, longitudinal layer is divided into three bands: the taenia coli

divided into a cortex and medulla: the lymphatic nodules occur in the cortex only. The lymph node is a filter of lymph: lymph enters at the periphery of the node *via* afferent lymphatics and flows through the sub-capsular sinus, to the sinuses of the connective tissue septae, through the lymphatic vessels of the medulla to the efferent lymphatic vessels that leave the gland at the hilum. Examples of these are shown in the top panel of Figure 4.5.

Spleen. Lymphatic nodules are present and pale centres occur as shown in the bottom right panel of Figure 4.5. There is no sign of a cortex or medulla. The spleen is a filter of both blood and lymph: a great deal more blood (red cells) is seen in the sinusoids of the spleen than in the inter-nodular tissue of the lymph node or in any part of the thymus. In fact, formalin pigment tends to form in the spleen because of the amount of blood present: see Chapter 5 for more details.

Thymus. NO lymphatic nodules are present. Tissue is arranged in lobes: each with a cortex and medulla, as shown in the bottom left panel of Figure 4.5. There are more lymphocytes in the cortex than in the medulla. The thymus has nothing to do with filtering lymph or blood. Unlike all other lymphatic organs, it is a primary lymphatic organ (the others, lymph node, tonsil, Payer's patches, mucosa-associated lymphatic tissue (MALT) and spleen, are all secondary lymphatic organs) and is concerned with the production of T lymphocytes.

4.4.2   Is the Tissue Normal?

Let us assume that you have identified the tissue either by applying knowledge and logical reasoning or by the simpler method of looking at the label on the slide. The next step is to decide whether the tissue

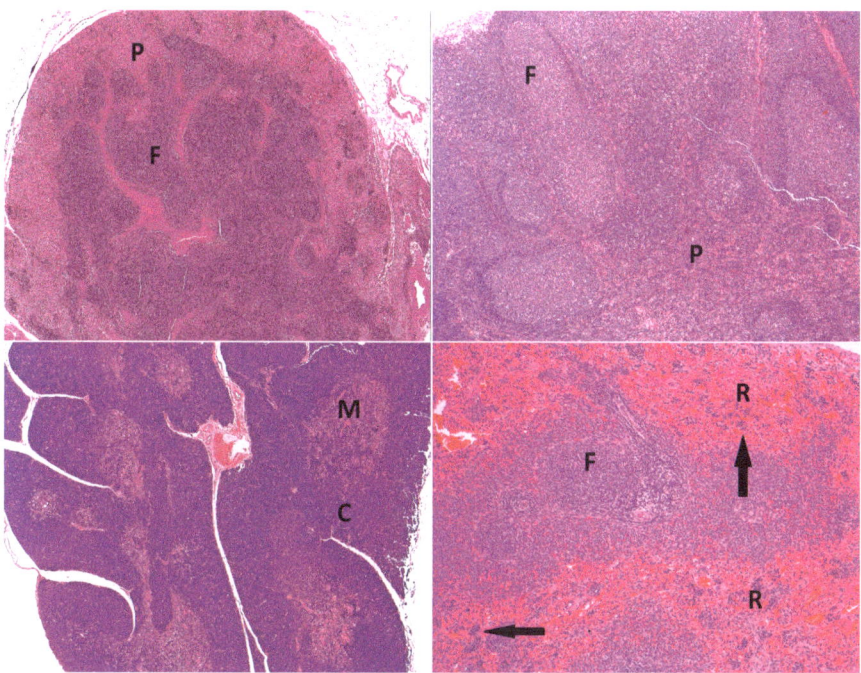

Figure 4.5 **Lymphoid tissue.** Top left is the pig lymph node—note the inside-out appearance with the parafollicular area (P) outside the follicular area (F); compare this to the more common appearance in the mesenteric lymph node of a dog on the top right. In the bottom panel is the thymus (bottom left) and the spleen (bottom right) of the rat. Note the lobular structure of the thymus, each with a darker staining cortex (C) and a lighter medulla (M). In the spleen you can see the follicle (F) and peri-arteriolar lymphoid sheath are populated predominantly by lympho-cytes, and in the intervening areas of red pulp (R) there are little clusters of very darkly staining nuclei (arrows). These are erythroid precursors. This is known as extra-medullary haematopoiesis and is a normal feature in young rodents, but may also be seen in some disease states.

is normal and, if not, what pathological processes have been in play. Again, it is essential to begin at low power.

4.4.2.1 Architecture. Knowing the architecture of normal tissue and comparing it mentally with the appearance of the section is the first and most important step. If you are not familiar with the normal appearance of the tissue, examine a control section. This control section should, of course, have been processed in exactly the same way as the section on which you are trying to make a decision. In addition, it should have come from the same part of the organ concerned. The control section will not, of course, be identical with what we might call the "test section", but it should present a very similar appearance if the test section is normal.

It is important to decide, in advance, the features of the tissue that you intend to assess when deciding on the normality, or otherwise, of the architecture. This might be called an exercise in pattern recognition: interestingly, some people are rather better at this than others. Learning to be good at it requires practice and an organised approach. If looking at a section of the wall of a large airway, for example, one might think in terms of:

- The layers present: are all the usual layers present? Are they of normal thickness?
- The major features: submucosal glands: are they present? Are they of normal size?
- The cartilage: Is the structure normal?

Or, thinking of a section of skin:

- Are both epidermis and dermis present?
- Is the epidermis of normal thickness?
- Is there a clear dividing line between the epidermis and the dermis?
- Are dermal papillae present?
- Are these of normal size and shape?

One of the most important features to examine when looking at the architecture of tissues is the way in which layers of cells lie together, or are they separated by what appear to be gaps? Oedema fluid, from leaky capillaries involved in an inflammatory response (see Chapter 2), pushes tissues and cells apart. But such separation can also be an artefact, especially an artefact of processing. Recognising such artefactual change takes experience: reference to control sections processed in exactly the same way as the section being examined is essential.

Drawing up a grid or table of the features you intend to look at before you look at the section is an excellent idea.

4.4.2.2 Architecture at Higher Magnification. Now we will be looking at the types of cells present, at the variety of cells present and at how they are arranged. We shall also be looking at the connective tissue: collagen, elastic fibres and reticulin fibres. The need for special stains will be apparent. First, look at the types of cells present.

Imagine an area of connective tissue, perhaps the connective tissue of a portal tract of the liver. You will have already decided whether the portal tract is of normal size in relation to the rest of the tissue. Has it expanded? Is it encroaching on the liver lobules?

Draw up a list of what you expect to see. You should expect to see fibroblasts, *i.e.*, elongated cells with elongated nuclei. The cytoplasm is difficult to see but the rather elongated nuclei are easy to find. How densely are they packed into the tissue? Look for other cellular components: macrophages (larger nuclei, cytoplasm extending out from the nuclear zone in a number of directions—not all that different from fibroblasts at first glance—perhaps containing pigmented material that might be brown or black in colour); lymphocytes with more nucleus than cytoplasm, the nuclei being darkly stained; polymorphs with their faintly granular cytoplasm (eosinophils with their bright pink granules) and lobed nuclei; mast cells with their dark nuclei and cytoplasmic granules (much easier to find if a special stain, for example, thionin blue, has been used because the granules are metachromatic); plasma cells with their large nuclei with the chromatin arranged in spokes, or peripherally placed lumps and a small pale area in the cytoplasm adjacent to the nucleus (the Golgi apparatus); and fat cells that, in ordinary paraffin wax sections, look empty—and so they are because the fat has been dissolved during processing. Forming an impression of the numbers of these cells present in a normal portal tract is important. Of course, you don't need to count them, but some impression of the normal density (cells per unit area of tissue in the section) is important. Even in "normal" sections you will see areas of abnormality. Small patches of densely packed mononuclear cells are often seen in the portal tracts of laboratory animals. Similar patches are seen near the airways of the lung. There, these might represent a low grade, but long standing infection. Gaining an impression of the limits of normality is very important: the pictures in atlases of histology are usually chosen to represent the ideal structure of the tissue, but departures from the ideal are common in real life.

Now turn to the connective tissue fibres. Collagen is stained pink in an H and E section and arranged as wavy bundles. Special stains will make the analysis of the connective tissue fibres much easier. Is there a lot of collagen present? How would you know? This could be known only by comparison with a normal section. Elastic fibres are seldom increased in amount, but reticulin fibres certainly are in, for example, the early stages of fibrosis.

4.4.2.3   Details of the Cells.   Perhaps the most important feature of a well-organised tissue, such as an epithelium, is the way in which the cells are organised. Are they in layers? Are the cells orientated in a consistent way? Examine a section of skin and look at the basal layers of the epidermis under high power. Note that the basal cells are orientated with their long axes vertical to the basement membrane. Note that they form a neat layer: there is no jumbling of the cells. Note that occasional mitoses can be seen. You will have to search to find mitoses: you will scan a number of high-power fields before you convince yourself that they occur! Note that the basal layer sits neatly on the basement membrane. No basal cells are seen below the basement membrane. Now look at the nuclei. Are they all much the same in appearance with roughly the same intensity of staining? Are they roughly the same size? Do they have roughly the same pattern of distribution of chromatin? Or are there areas in which the nuclei are pale and enlarged? Are there small, dark nuclei? You will see some in any normal section, but not many.

Again you might draw up a grid or table of the features of the cells you intend to assess.

All this might seem a trifle laboured and indeed it is. But this is the process followed by experienced histologists, although they themselves might not recognise it. "I can tell, at a glance, if a section is normal!" "Yes, no doubt, but that simply shows how quick you are at assessing the key features of normality!" Acquiring this skill takes practice. Drawing up a grid and looking at the features you have identified in an orderly way is like practising scales on a piano: essential if you intend to play well! Let us take another example.

4.4.2.4   Looking at Liver.   Let us imagine that we are examining a section of liver: something toxicologists often need to do. You should remind yourself of the key features by examining a control section.

1. Examining the control section
   The architecture is arranged in lobules. The "classical lobules" are hexagonal, centred on a tributary of the hepatic vein and bounded by fine connective tissue that contains the usual elements of loose connective tissue, as shown in Figure 4.6. If you are looking at tissue from a pig (or camel, though that is unlikely), the lobules will be easy to see: the connective tissue that defines the lobules is unusually well-developed in this species. Unfortunately, it is not as well-developed in other species, including man. Remember that the section will cut through the lobules at random.

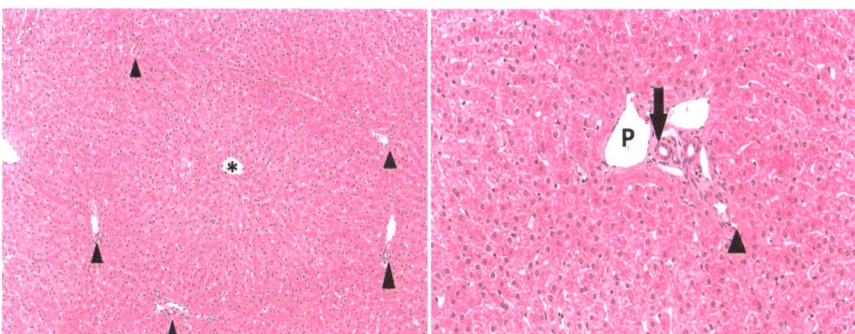

Figure 4.6 **The liver of the pig is probably best for visualising structure.** On the left you can see the central vein (asterisk) surrounded by five portal triads (arrowheads). In the right hand higher power image you can see the bile ducts in the centre (arrow) with their cuboidal lining, and at about five o'clock there is a branch of the hepatic artery (arrowhead). The other vessels are the portal vein branches (P). You may be able to see a little brown bile pigment if you study the image closely.

Some sections of lobules will appear as small areas in comparison with others. This should make one think about the three dimensional shape of the lobules: are they all 18-sided? In all species the centre of the classical lobule is a tributary of the hepatic vein known as the central vein. Having found a central vein, scan outwards towards the periphery of the lobule looking for portal tracts. These contain the three-component systems that run at the edges of the classical lobule. Each portal triad comprises a branch of the hepatic artery, a branch of the portal vein (a branch, not a tributary: blood is flowing towards the liver through the portal vein) and a small bile duct. Finding the bile duct is easy: it is the only small vessel in the triad with a lining of cuboidal epithelium. The cuboidal epithelium often looks superbly regular: perfectly aligned nuclei, all cells the same height: it really does look cuboidal. Not all "triads" contain three vessels. Here and there, only an artery and a bile duct, or perhaps only a vein and bile duct, will be seen. This should make one wonder about just how regular the classical hexagonal structure actually is. Note that the bile ducts are not always cut at a perfect 90° to their long axes: remember that sectioning is random. Note that the vessels of the portal tract are surrounded by loose connective tissue. Lymphocytes and plasma cells are seen in this loose connective tissue (see above). Here and there, a space may be seen. This might represent an artefact (shrinkage during processing in paraffin wax) or a small lymphatic vessel. All that has been described so far can be seen under low power (10× objective).

Now look back inwards towards the central vein. Move to higher power and look closely at the hepatocytes. These seem to be arranged in lines, plates in three dimensions. The plates are actually double layers that make up the muralium of the liver. We shall be concerned to look at the three areas that often show changes as a result of exposure to toxic chemicals: the peri-portal tissue, the hepatocytes around the central vein and the ones in between. Centrilobular necrosis is the commonest finding: the cells around the central vein, and for some distance out into the lobule, may be necrotic and show fatty degeneration. This will not, of course, be present in the control section.

Now examine the portal tracts and the fine connective tissue around the periphery of the lobule under higher power. Form an impression of how much connective tissue is present. If you have a section that has been stained for collagen this will be much easier. The reason for doing this is that we will be looking for abnormal fibrosis in the test section and need a baseline for comparison. Look again at the vessels of the portal triads: take a closer look at that cuboidal epithelium of the bile ducts. If you scan around the section you may come across an area of connective tissue that contains a large bile duct. Goblet cells appear amongst the cells of the cuboidal epithelium

2. Examining the test section

   Scan at low power. Does anything "jump out" at you? For example, there might well be present a large metastasis from some primary tumour of some other tissue. The reason it "jumps out" is that it is not liver. Metastases to liver are usually round in shape and sections through them often appear as circular areas. Look next at the distribution of connective tissue. Is this normal or are the lobules "caught" between thick bands of connective tissue? The connective tissue will, of course, be much easier to assess if a collagen stain has been used (see Chapter 7). In advanced cirrhosis the lobules are small islands surrounded by seas of connective tissue. Look at the portal tracts and at the bile ducts. Are there several, or many, transections of bile ducts seen in one portal tract? Bile duct hyperplasia is a common response to toxicological agents.

   In examining the section of liver you should have the control section alongside the test section. Switching back and forth between the two is illuminating. Of course, no experienced pathologist does this, he rests confident in his knowledge of what the normal tissue looks like.

Going further into the functional architecture of the liver

No self-respecting hepatologist speaks of classical lobules. Kiernan described the classical hexagonal lobule, which we discussed above, in 1883. Note that the lobule is a solid structure and cannot, therefore, be correctly described as hexagonal in shape. The transections might be hexagonal in shape and often are. Mall[†] (1906) pointed out that an alternative hexagonal lobule centred on the portal tract might be a better description: this was the portal lobule. And there the matter rested until 1954, when Rappaport revolutionised thinking about the structure of the liver. Not much has happened with regard to how we think about the structure of the liver since then. Rappaport's paper (A M Rappaport, Z J Borowy, W M Logheed and W N Lotto, Subdivision of hexagonal liver lobules into a structural and functional unit, *The Anatomical Record*, 1954, **119**, 11–27 plus three plates), is so good it should be read by anybody interested in the liver. The illustrations are splendid. In essence he proposed an entirely new concept: small acini radiating outwards from the portal tracts and each divisible into three zones in terms of their oxygenation, which of course depended on their blood supply. The zone nearest the portal tract was well-oxygenated, it received a lot of its blood from branches of the hepatic artery; the more peripheral zones were less well-oxygenated. The direction of flow of the blood was, of course, from the portal tracts to the central veins; bile flowed through the cholangioles between the hepatocytes of the muralium to the small canals of Herring at the periphery of the classical lobule and into the bile ducts. Rappaport's scheme explained the "bridging" of damage that is seen in some types of liver damage: not all the damage is seen in the cells around the central veins; bands of damaged cells extend out towards the periphery of the classical lobule and meet the margin of the classical lobule about half way along a "side" between the portal tracts. Rappaport also pointed out that the arrangement of the acini was irregular, another departure from the orderliness of the classical concept. His description is worth repeating:

> *"The newly defined structural unit is a small, irregular, berry-like parenchymal mass situated around the trio of the terminal branches of portal vein, hepatic artery and bile duct, growing out from a small portal triad and mainly running perpendicularly to the central vein. The hepatic unit occupies adjacent parts of the neighbouring hexagonal fields and extends from the central vein of one hexagon to the central vein of another."*

[†]For details of liver structure, see histology textbooks listed in the bibliography.

The surprise in this is the observation that the functional unit, the acinus, extends across from one classical lobule to the next. One could imagine it as a spindle-shaped structure, with its short axis defined by the branches of the vessels in the portal tract and its points reaching the central veins of two classical lobules. For more details, see the original paper. Note that, in describing this new concept, Rappaport anchored his description on the hexagonal, classical architecture described by Kiernan. The key point is that the functional architecture of a tissue need not reflect the structural architecture. How surprising: form nearly always follows function. Very interestingly, pathological changes supply clues to the functional architecture of the system.

4.4.2.5 Conclusion. An organised approach to the examination of sections is essential. There is much to be said for drawing up a grid or table of features that you intend to examine and assess. In each case the key is comparison with the normal: sections of normal tissue prepared in precisely the same way as sections of test material should always be examined before and alongside the test sections. Preparing "new" control sections is essential: if old sections are used, one cannot be sure that the method of preparation was identical with that of the test sections. This is a "counsel of perfection": in a laboratory doing a lot of routine work there is usually great consistency from one batch of sections to the next. But, if quantitative work is to be undertaken, there is no substitute for "new controls".

4.4.3 The Morphological Method

The morphological method involves making inferences based on an appreciation and analysis of structure. These inferences may relate to the development of structure, of how it has evolved and about how it functions. To the pathologist, recognition of damage and change are the key observations. From these, inferences about the effects on the whole organism can often be drawn. Inferences regarding mechanisms of injury often rely on recognising patterns of change in structure that are similar to those for which mechanistic explanations have been worked out. The morphological approach can often be combined with a study of function. Labelling thymidine and following its appearance in cells over a period of time provided key information on cell division.

All morphological studies depend on being able to define and recognise structures. Recognition and definition are linked: recognition depends on definition: you have to be able to define a structure before you can recognise it or distinguish it from other structures.

Definition involves description. Description should be as detailed and complete as possible: the finer the description, the better the definition. Description of structure involves a verbal report of the appearances of the structure. One can only describe what one can see, and thus description depends on what is visible. What is visible, in turn, depends on the techniques applied. Nobody can describe a mitochondrion with the aid of only a magnifying glass; more can be done with the light microscope and a lot can be done with the electron microscope. How much one sees of a visible structure depends on one's powers of observation. These depend on how carefully you look and on how hard you think as you look. The beginner sees little on taking a casual glance at a section; the histologist sees a great deal on examining it carefully. Training one's powers of observation is important for anybody doing morphological research. This can be done by reading the descriptions provided by experts and then looking for yourself and discovering how much you can see.

4.5   DRAWING (FOR ENTHUSIASTS ONLY)

Drawing used to be required in histology practical classes in all universities; it is no longer required. It is sometimes said that photography has replaced drawing; this is only partly true. Of course, the photograph provides a better and perhaps more accurate picture of the section than a drawing, but it fails to provide the stimulus to look critically that is provided by drawing. In examining a photograph one is faced by the same problem as faced in examining the section: one's powers of observation are, again, tested. Of course, if the photograph is labelled and all the important features pointed out, the problem is, to some extent, solved. In research the photographs are not labelled any more than the sections are labelled, unless you do it yourself. Not many readers will have been convinced by this and will be losing interest. For the few enthusiasts still reading, I shall say a little more.

Drawing forces you to notice things; that is its greatest virtue. Drawing is a valuable part of the process of studying a section, drawing for publication is more demanding and few would now aspire to this. Could anybody produce a drawing that was as useful as a colour photograph? Well, perhaps not now, but anybody who has seen Maximow's drawings that were used to illustrate his textbook (later Bloom and Fawcett: see Bibliography) will understand that this used to be possible. Not many, perhaps very few, can draw like Maximow, but everybody can use drawings to help them examine histological sections. Drawing is particularly helpful in exploring spatial relationships, indeed changes in spatial relationships are difficult for many of us to grasp without "doing a drawing".

4.5.1   Drawing as an Aid to Examining Histological Sections: How to do it

1. Distinguish between a drawing and a diagram. A drawing is a faithful representation of that which can be seen. There is almost no limit to the detail that can be included in a drawing; one is limited, in fact, only by visual acuity and manual skill: see Maximow's drawings. The realist school of art has carried this process to its limit. A diagram is different: it illustrates spatial relationships, often in outline, and does not purport to reproduce all the visible features of the object. The most useful form of illustration for the histologist is what might be called the "developed diagram". Here, the outline is included but certain features are explored in more detail.

2. Scale. Always draw to scale. Using an eyepiece micrometer that has been calibrated against a stage micrometer is the easiest way of keeping to scale. Begin, always, by drawing a scale line on the paper. Measure structures.

3. Now the tricky bit. Draw the outline first. Never begin with an obsession for reproducing the pattern of chromatin in a nucleus. Even if you are interested in only one cell, draw the cell first, then the nucleus, then the chromatin. At this point, many will say, "but I cannot draw!" Do not lose heart! Some fortunate people can sketch what they see: this is a talent. If you have it, be grateful; if you don't, read on. Draw a rectangular grid on the paper. Draw the grid to scale so that you know the "real distance" between the lines. Now, working from square to square, draw in the outline. Don't make the squares of the grid too small, this does not help. A centimetre grid is about right. If you have an eyepiece graticule ruled as a grid, it will help but it is not essential. Matching the eyepiece grid to the grid on the paper is the ideal method. The grid you have drawn on the paper provides the cues for spatial relationships that those with a natural talent for sketching seem to possess as part of their neural processing of visual images. Having filled in the outline, not just the edges of the area you are interested in but also the outlines of the important features within that area, add some detail in a few smaller areas. The developed diagram is mainly an outline with a few areas of detail.

4. Aids to drawing. The camera lucida projects an image of the section on to paper: you draw on the paper, "on" the image. Very useful indeed, but there are not many of these devices in service, though they are still available from manufacturers of microscopes. If you have access to one, try using it. I have used this device but found the temptation of trying to fill in too much detail most distracting. Those interested in the "old days" will be interested to know that histologists used to draw by looking into a monocular microscope, keeping both eyes open, and fusing the images of the section and a sheet of paper placed on the bench alongside the microscope. Both the paper and the secondary image produced by the microscope, which you will be looking at, are about the same distance from the eye. Yes, not easy! Most right-handed people want to use their right eye at the microscope and their right hand for drawing; this is not possible. At least at first the images of the section and the paper, or the point of the pencil, keep sliding apart or obstinately refusing to fuse. Only historically minded enthusiasts need try this.

5. Tools (for those who have never drawn). Draw using a draftsman's clutch pencil, the Pentel® 0.5 mm lead, for example, is excellent; others are no doubt as good. Obtain H-grade leads; the usual HB leads are too soft. A well-sharpened H pencil is as good, but needs to be kept sharp: the finer side of an emery board should be used for renewing the fine point of the pencil between sharpening. Draw on good-quality paper. Use an artist's "putty" eraser. The old advice still holds: don't use shading, stipple if you must and make every line mean something. Colour? Unnecessary for a developed diagram.

6. Finally, for the small number of people who have got this far, drawing is a useful aid to examining sections. Try it and see.

4.6 ARTEFACTS

Artefacts are a perpetual problem in histological research on the toxicological effects of chemicals. Changes to what is regarded as the normal appearance of tissues could be caused by the chemical under study; they might also be caused by the chemicals to which tissues are exposed during processing or by an interaction between the former and the latter. Processing can also cause changes produced by the chemical being studied to disappear. Imagine, for example, that the effects of some chemical on mitochondria were being studied; it would clearly be important not to use a fixative that either destroys mitochondria or prevents them being seem on histological examination. Similarly, if the microvilli of the renal proximal tubule are being studied, it will be very important to process the tissue in such a way that these structures are not damaged by anything other than the chemical under examination.

Let us divide artefacts into two groups:

1. Positive artefacts: changes that appear as a result of processing of the tissue, which are not caused by exposure to the chemical being studied;
2. Negative artefacts: the disappearance of changes caused by the chemical being studied.

Positive artefacts can be dealt with by the use of negative controls. We know that NOT treating an animal with compound X cannot produce effects due to exposure to compound X. When the control tissue is processed and examined two possibilities appear:

1. The tissue may show changes from normal: these are positive artefacts produced by processing;
2. The tissue may appear normal.

If the tissue from the negative control does appear normal then we can delete the possibility of positive artefacts from the sections of tissue from the animals exposed to compound X. At least this is often assumed to be so. The deduction is, however, possibly false. Imagine that the presence of compound X in the tissue, whilst producing no effects itself, led to compound X combining in some way with the chemicals used during processing to produce the observed effects. The negative control does not invalidate this suggestion. A larger exposure might produce a larger effect and this might be interpreted as showing that compound X, per se, is having some effect, but this is also not certain. A range of exposures that produces a range of responses, larger effects at larger exposure, provides good, but not absolute, evidence of a real effect.

Processing the tissue in a variety of ways, for example, by varying the fixatives and staining methods, will help to invalidate the possibility of a positive artefact: it is unlikely (though not, in principle, impossible) that compound X would interact with all the chemicals that could be used during processing to produce the observed effects. Of course, the negative controls would also have to be exposed to the same variety of processes to provide assurance that they remain negative. Using a completely different method of processing would also help: frozen sections prepared from tissue exposed to compound X but not to any chemical fixatives and from tissue exposed to the fixatives but not to compound X would provide satisfactory controls. The need to use frozen sections is obvious: paraffin sections cannot be prepared without hardening of the tissue by fixatives. Exploring the effects of using and not using fixatives in the preparation of frozen sections is also a sensible suggestion. Using frozen sections eliminates the possibility, however remote, that heating the tissue to the melting point of the wax led to compound X producing the observed effects.

If changes, positive artefacts, are seen in the "negative controls", and these are found to be identical with the changes seen in the tissue exposed to compound X, it will clearly be impossible to safely interpret similar changes seen in the tissue exposed to compound X as being caused by compound X. Very close examination to establish whether the changes really are the same in the exposed tissue and in the negative controls will be necessary. If this is the case then a range of methods of processing should be tried until one producing no effects in the negative controls is found.

The negative artefact cannot be eliminated by negative controls. Non-appearance of the change in the tissue exposed to compound X could be a result of the change not being there (as in the negative control), but also as a result of the change being destroyed by processing. Positive controls might be suggested, but these are impossible in the face of a suggested negative artefact. When searching for a no-effect level of exposure to a compound, the problem of the negative artefact becomes acute. Excluding the possibility of a negative artefact at low levels of exposure, whilst accepting that there is an effect at higher levels of exposure, is impossible. Using any other chemical known to produce the observed effects is clearly illogical (the "other chemical" and compound X might work in different ways), though using a related compound (something chemically similar to compound X) might be helpful. Once again, only the use of an entirely different processing method will eliminate the possible effects of processing.

A wide range of artefacts may be produced by the normal methods of processing tissue for the preparation of paraffin wax sections. These should be understood before sections are interpreted. Artefacts that can be produced during processing of tissue are discussed in Chapter 5. Control for artefacts is critically important in histochemical methods and will be discussed in Chapter 9.

Tissue Processing: Fixation, Dehydration and Clearing

Processing specimens of tissue from post-mortem sampling to finished histological sections is, of course, a critical part of the practice of histology. Often, this aspect of the work is left to technicians, who do it very well indeed. But the research worker might well wish to process his or her own tissue or, at the very least, to understand what has been done to it en route, so to speak, from the post-mortem bench to the stage of the microscope. In this chapter the basics of tissue processing are explained. We begin with fixation, a critical step. Then, we consider preparing the tissue for embedding in a supporting medium ready for sectioning. The medium considered in the next chapter is paraffin wax—still perhaps the most widely used of all the media available—and so the methods discussed here are suitable for preparing tissues for this form of embedding. Many of the techniques described in this chapter and the next are well over one hundred years old. This does not mean that they are outdated; on the contrary, it speaks of their remarkable reliability!

5.1 FIXATION OF TISSUE

Before tissue can be processed for sectioning in paraffin wax or celloidin, it must be fixed. Only frozen sections can be cut from unfixed material. Fixation is also essential for the preparation of sections for

Histological Techniques: An Introduction for Beginners in Toxicology
By Robert Maynard, Noel Downes and Brenda Finney
© Maynard, Downes and Finney 2020
Published by the Royal Society of Chemistry, www.rsc.org

electron microscopy. Fixation is perhaps the most critical stage of tissue processing: faults occurring during fixation cannot be corrected later. In this, fixation differs from staining: faults occurring during staining can often be rectified. Despite this, little attention tends to be paid to fixation and the almost universal use of formaldehyde, usually as "10% buffered formalin" (see section 5.6.1) makes some workers wonder whether anything much needs to be said. A glance at any of the manuals of histological technique will show that many dozens of different fixatives have been developed: Gray (see Bibliography) listed about 500 different mixtures, although, admittedly, some of these were not very different from one another. The names of the inventors of some of these fixatives will be familiar to most histologists even if they use, almost exclusively, formaldehyde: Bouin, Zenker, Heidenhain, Carnoy. These fixatives are often referred to by their authors' names: "fix in Zenker's fixative" or just "fix in Zenker". To some extent, the choice of fixative is a matter of tradition or habit. Histologists become used to a particular fixative and adept at examining sections of tissue fixed in a particular way. Such habits vary from country to country: for many years Bouin was the most widely used fixative in France and Zenker's fixative was widely used in the United States. Unsurprisingly, the inventors of fixatives advocate use of their own inventions. For routine work, formaldehyde produces good results; but, for research work, it is at least worth exploring other fixatives and using those that produce the best results.

In some laboratories the use of formaldehyde has been abandoned and commercial fixative mixtures known by their trade names are in general use. These are often based on the dialdehyde glyoxal and work in much the same way as formaldehyde. The exact composition of these mixtures seems curiously difficult to determine: patents have been sought by and granted to their manufacturers. Dialdehyde mixtures are sometimes seen as a "safe" alternative to formaldehyde. This is largely on the grounds that they have a low vapor pressure and are said to be biodegradable, and thus problems of disposal are reduced. They are, however, skin irritants and, like formaldehyde, may produce sensitisation. They should not be regarded as "harmless". Recall that any effective fixative will fix the skin of your fingers if exposure is prolonged. In general they produce good fixation and they are also useful if immuno-staining techniques are to be used. Detailed studies comparing these commercial mixtures with a range of classical fixative mixtures appear to be limited. There seems, to the present author, to be something unsatisfactory about using a fixative mixture that has not been compared with other fixatives and about the composition of which one is unsure.

Concerns similar to those expressed about formaldehyde (see section 5.6.1) have led to a number of classical fixative mixtures falling into disuse on the grounds of Health and Safety. Such concerns should be taken seriously. Mercuric chloride is both corrosive and poisonous and needs to be disposed of so as not to allow contamination of the environment with mercury; picric acid, when dry, can be detonated. That these compounds can be handled safely is, however, certainly true. Indeed, much more dangerous chemicals are handled in all chemistry laboratories. Abandoning the use of mercuric chloride on the grounds that it is simply too dangerous to be used is an over-reaction and research workers should make their decisions about fixatives on better, more scientific grounds. The sensible thing to do is to experiment with a range of fixatives until one finds the optimal mixture for use in one's own work. Such experimentation requires the ability to distinguish between the effects of different fixatives. Advice on this is provided in the following sections.

5.2   OBJECTIVES OF FIXATION

Fixation has essentially two objectives:

1. To preserve tissue, including, of course, the cells that comprise the tissue, in a state as similar to their living state as possible.
2. To protect tissues from damage that might occur during the processing to sections, whether they be of paraffin wax or celloidin or, in the case for electron microscopy, of some type of resin.

In addition, fixation may usefully alter tissues in such a way that staining with certain dyes is enhanced.

5.2.1   Preservation of Tissue

In preserving the structure of tissues, the effects of drying, dissolution and decay must be prevented. A piece of liver left on the bench will shrink and dry and, eventually, rot. The decay of the tissue is in part due to autolysis by enzymes released from lysosomes as cells break down and in part due to colonisation by bacteria and molds. If a piece of liver is placed in a beaker of distilled water it will soon fall apart: the cells will take up water and burst. Fluids that prevent these processes are called preservatives.

Fixatives do all that preservatives do but, in addition, harden tissue so that it will not be damaged by dehydration and exposure to embedding media.

5.2.2 Protection of Tissue from the Effects of Dehydration and Embedding

Dehydration of fresh tissue damages cells. Imagine placing a piece of liver in a concentrated solution of sodium chloride: it would soon lose water and the cells would shrink rapidly. Exposure to absolute alcohol would produce similar results. Even with a gentler process of dehydration it is likely that soluble substances would be moved about within the tissue as water was extracted. In processing for embedding in paraffin wax all the water must be removed from cells and replaced with a solvent that is miscible with paraffin wax. Dehydration in a series of mixtures of increasing concentration of alcohol and water (beginning with 50 or 70% alcohol and increasing to 100% alcohol) is followed by soaking in a solvent such as toluene. The water of the cells is miscible with alcohol, the alcohol is miscible with toluene and the toluene is miscible with paraffin wax. Substances soluble in water are in danger of being washed out during the dehydration stages; lipids will, of course, be dissolved by the alcohol. When processing is complete, all the "space" that had been occupied by water and its solutes is replaced by paraffin wax: every cell is filled, or infiltrated, with wax. Those substances that were insoluble in water or that were rendered insoluble in water or any of the other solvents used in the process by fixation will remain. That they should remain where they originally were is critically important. Fixation must prevent, as far as possible, shrinkage of the tissue during dehydration and exposure to molten paraffin wax. Some shrinkage is, in fact, inevitable, but effective fixatives, whilst not preventing all shrinkage, do prevent distortion during shrinkage: this is important. Some material that would otherwise be removed during processing can be retained by fixation. Baker's calcium-formalin fixative stabilises lipids, especially phospholipids, and allows them to be stained. Fixatives containing chromium trioxide stabilise glycogen and, similarly, allow it to be stained.

The main effect of fixatives is to alter proteins in such a way as to render them resistant to destruction and distortion during the remainder of processing. This is achieved by either binding to the proteins or affecting them in such a way as to alter their interaction with water. In some cases the proteins are coagulated, as egg white is coagulated by heat; in others no visible coagulation occurs but the proteins are in some way stabilised.

5.3 CHEMICALS USED AS FIXATIVES

Relatively few chemicals are used as histological fixatives. These few compounds, the primary fixatives, have been combined to give the large number of fixative mixtures mentioned above. With the exception of formaldehyde, no primary fixative is used alone. Mixtures have been produced to provide properties not provided by the primary fixatives. For example, some produce swelling, others produce shrinkage, but when combined these effects offset each other and the other, often very desirable properties of the primary fixatives are retained. The best account of fixatives has been provided by Baker (1958): he devoted 150 pages to the subject. Much of his discussion centered on mechanisms of effect, especially of primary fixatives. He devoted almost no space to a discussion of fixative mixtures. Here, we shall reverse the emphasis and simple list the primary fixatives, describe briefly how they work and focus on the fixative mixtures that a research worker might wish to consider. A close reading of Baker's account reveals that much remains to be discovered about how fixatives actually work; even less well understood is how they work when combined.

5.4 THE MECHANISMS OF FIXATION

All fixatives affect protein. Some cause solutions of proteins to coagulate, *i.e.*, to form a clot. Proteins occur in tissues as gels rather than as solutions: fixatives stiffen protein gels. Fixatives can also change the relationship between protein and water in a gel. Consider a gelatine gel: no amount of pressure will squeeze water from the gel. But if the gel is placed in formalin it becomes tough and fibrous (fixed) and water can be squeezed from it as from a sponge. But formalin does not cause coagulation of a solution of albumin. These properties led to the classification of fixatives as coagulants and non-coagulants. Formaldehyde is a non-coagulant. It is for this reason that it is less effective than, for example, mercuric chloride in preventing shrinkage during dehydration and embedding in paraffin wax.

A second division into those that react with protein and produce chemical bonds between chains of amino acids and those which do not, led to fixatives being classified as additive and non-additive. Rather oddly, formaldehyde is very effective at forming cross links

Table 5.1 Classification of primary fixatives.

Primary fixative	Non-additive	Additive	Coagulant	Non-coagulant
Methanol	Yes		Yes	
Ethanol	Yes		Yes	
Acetone	Yes		Yes	
Nitric acid	Yes		Yes	
Hydrochloric acid	Yes		Yes	
Acetic acid	Yes			Yes
Potassium dichromate in alkaline solutions	Yes			Yes
Mercuric chloride		Yes	Yes	
Chromium trioxide		Yes	Yes	
Picric acid		Yes	Yes	
Formaldehyde		Yes		Yes
Osmium tetroxide		Yes		Yes

between chains of amino acids, but does not produce coagulation. Baker could not explain this; nor can I. Ethanol is a coagulant fixative but certainly does not bind to amino acids. Primary fixatives that do not bind to amino acid chains are also referred to as denaturing fixatives. Table 5.1 sums up the classification of the major primary fixatives.

It will be apparent that most primary fixatives produce coagulation of proteins and that some, both additive and non-additive, do not. It will be realised that some primary fixatives are reducing agents, others are oxidising agents. Rather remarkably, formaldehyde can be mixed with potassium dichromate to produce a very valuable fixative mixture. Such mixtures are sometimes described as irrational, and so they are if we consider only the redox properties of their components.

5.5 OTHER PROPERTIES OF PRIMARY FIXATIVES

5.5.1 Penetration

An ideal fixative would penetrate tissues rapidly, fixing as it went. Some, like formaldehyde, do penetrate rapidly, though fixation lags behind penetration; others penetrate rather slowly but fix rapidly. The rate at which fixatives penetrate tissues conforms to Fick's Law of Diffusion. This can be expressed as shown in eqn 5.1:

$$d = K\sqrt{t}, \tag{5.1}$$

where d is the distance travelled, t is the time taken and K is a constant that varies from fixative to fixative.

If we take logarithms, we achieve eqn 5.2:

$$\log d = 0.5 \log t + \log K. \tag{5.2}$$

Thus, for all fixatives, a plot of log$d$ against log$t$ should produce a straight line with a gradient of 0.5. The intercept on the y axis will define log$K$.

Alternatively, if time (t) is plotted on the x axis and scaled as a series of squares and distance is plotted arithmetically on the y axis, a series of straight lines will be produced with their gradients being defined by K.

We have been considering Fick's First Law of Diffusion. This is expressed by eqn 5.3:

$$\frac{dm}{dt} = KA\left(\frac{dc}{dx}\right), \tag{5.3}$$

where $m =$ mass of material, $t =$ time, $A =$ area across which diffusion is occurring, $c =$ concentration and $x =$ distance.

In terms of dimensions:

$$MT^{-1} = k[L^2ML^{-3}L^{-1}] \tag{5.4}$$

$$T^{-1} = kL^{-2} \tag{5.5}$$

$$1/T = k(1/L^2) \tag{5.6}$$

$$L^2 = Tk \tag{5.7}$$

$$L = K\sqrt{T} \text{ (where } K = \sqrt{k}). \tag{5.8}$$

If d is measured in mm and t in hours, then K is the number of mm the fixative will have penetrated in one hour.

K values for a few primary fixatives are provided in Table 5.2.

Table 5.2 Primary fixatives: distance penetrated in one hour.

Primary fixative	K (mm)
Formaldehyde	3.6
Acetic acid	2.75
Mercuric chloride	2.2
Chromium trioxide	1.0
Osmium tetroxide	0.85
Picric acid	0.8

Table 5.3 Penetration of liver tissue by mercuric chloride.

Time	Distance
2 s	20 μm
1 h	0.84 mm
77 days	3.6 cm

The lesson from all this elementary mathematics is **TAKE THIN TISSUE BLOCKS!** If not yet convinced, consider Table 5.3 of the penetration of mercuric chloride.

In practice, no block thicker than about 5 mm should be taken. The other dimensions of the block are irrelevant as far as penetration by fixatives is concerned.

It should be noted that the figures quoted in Table 5.2 above, from Baker, come from studies of fixation of albumen–gelatine gels and should not be taken as figures for penetration into tissue. Note also that fixatives penetrate into soft tissue, like liver, a good deal more rapidly than they do into hard tissue, like tendon.

5.5.2 Swelling and Shrinking

Primary fixatives may produce swelling or shrinkage of tissue. As a rough guide:

- Acetone and ethanol produce marked shrinkage;
- Chromium trioxide, osmium tetroxide, mercuric chloride and picric acid produce little effect;
- Formaldehyde produces mild swelling of gels but not much change in the volume of a piece of liver;
- Acetic acid produces very marked swelling.

The shrinkage produced by ethanol can be offset by the swelling produced by acetic acid: Carnoy's fixative contains both these primary fixatives. Swelling and shrinkage have little to do with the osmotic pressure of the primary fixatives. 5% acetic acid produces very marked swelling, though its osmotic pressure is about three times that of plasma. If osmotic pressure is not important, one might wonder about the purpose of adjusting the osmotic pressure of the fixative mixture: why should formol-saline be any better than formalin in water? A number of salts have been added to fixative mixtures: sodium sulphate is often included in the formulae of Heidenhain's Susa mixture. Some authorities simply delete such additions; others regard them as important or, at least, as harmless. When fixing tissues for light microscopy, osmotic pressure does not seem to be important; for electron microscopy (EM), it is generally acknowledged to be very important and sucrose is added to EM fixatives.

Much of the above discussion about swelling and shrinking is, disappointingly, less useful than might have been hoped. What really matters is how much swelling or shrinking has occurred when processing has been completed and the sections have been cut. The general rule is that shrinkage always occurs, irrespective of the fixative used. Formaldehyde is one of the poorer fixatives for defending against shrinkage; mercuric chloride is, generally, more effective.

5.6 FIXATIVE MIXTURES

Fixative mixtures can be divided into two groups: those intended for micro-anatomical (meaning general histological) work and those for cytological work. There is some overlap between the two groups. We shall consider the micro-anatomical fixative mixtures in some detail, and then turn to the more specialised cytological fixatives.

As stated above, there seems little point is discussing primary fixatives because, with the exception of formaldehyde, they are always used in mixtures and the properties of the mixtures differ from those of their components. The exception, formaldehyde, will be discussed.

5.6.1 Formaldehyde (HCHO)

Formaldehyde is a gas that is soluble in water. A saturated aqueous solution of formaldehyde contains about 40 g formaldehyde per 100 mL solution (40% *w/v*). This concentrated solution is described as formalin. It is of no use as a fixative: it produces far too much hardening and shrinking of tissue. It is usually diluted with water (but see below) to produce a 4% solution of formaldehyde. Now for some confusion. This 4% solution is usually referred to as 10% formalin. There's nothing wrong with that: a 10% solution of a 40% solution is a 4% solution! Some authorities refer to it as 4% formaldehyde, others as 10% formalin. Solutions of formalin undergo oxidation and formic acid is produced. Such acid solutions inhibit staining with, for example, eosin, and the 10% formalin is usually neutralised. The classical method is to store the 10% formalin over a layer of marble chips. A better method is to buffer the solution with a phosphate buffer. Neutral-buffered formalin means a 10% solution that has been buffered. It is worth remembering that slight acidity is actually helpful for silver stains: see Chapter 7 for a discussion of silver impregnation methods. Drury and Wallington (see Bibliography) point out that marble chips should not be added to 40% formalin: the carbon dioxide produced can cause the bottle to explode.

Formaldehyde is an irritant, a carcinogen and sensitising agent: it deserves to be treated with respect. The vapour is an irritant and formaldehyde should not be used on the open bench: use a fume hood or fume cupboard. Gloves are essential. Many workers, including the present author, used to ignore this advice and developed dermatitis. This was unpleasant: the author's eyelids used to swell and desquamate. All this disappeared as soon as a fume hood and gloves were introduced. Rather oddly, some workers appear immune from the effects of formaldehyde and dip their hands in the 10% formalin fixative for years with no obvious ill effects. Tempting fate in this way is not advisable!

5.6.1.1 Use of 10% Buffered Formalin as a Fixative. 10% buffered formalin (often referred to as 10% formalin) is a very widely used fixative: some pathologists use nothing else. Such wide use suggests that it has few drawbacks and, indeed, this is true. A few faults are worth mentioning. Firstly, formalin does not stabilise tissue as well as other fixatives and shrinkage during further processing is often produced. Using a second fixative, see 5.6.2, can prevent this. Secondly, if tissue containing a lot of blood is exposed to formalin, a black pigment resulting from the combination of formaldehyde with haematin (from haemoglobin) is produced. This can be removed from sections but the additional steps are a nuisance. Thirdly, now and again tissue fixed in formalin fails to stain well with haematoxylin and sections appear pink rather than pink and blue. This is referred to as "pink disease". Ensuring complete dewaxing of sections and treating sections with 1% HCl "cures" the disease. Some authorities have argued that tissue fixed in 10% formalin is rather too soft for perfect sectioning after embedding in paraffin wax. Well, mercuric chloride-based fixatives probably do produce blocks that cut a little better but, given the number of sections cut from tissue fixed in 10% formalin per year, this cannot be a major drawback.

10% formalin has a number of advantages: it is easy to use, it is cheap, making up a complicated mixture is unnecessary, tissue can be left in the fixative for long periods and still processed satisfactorily and it is as good a fixative for storing tissue in as any. These are not negligible attributes.

Gabe was less restrained than many about the qualities of 10% formalin:

"It is one of the poorest preparations for paraffin embedding as it is for most so-called topographical stains. Only a combination of in-difference and slavery to routine can account for the considerable vogue amongst anatomopatholgists for fixation in formalin without any adjuvant."

In speaking of adjuvants Gabe mentioned adding saline to the solution (formol-saline) and calcium salts (*e.g.*, Baker's calcium formol and Lillie's calcium acetate formol) but remained very critical of this fixative. He provided a list of four fixative mixtures that he recommended trying when a tissue new to the research worker involved was being studied. Gabe was French, and Bouin's fixative was the standard fixative in French laboratories for many years. Gabe did not, however, spare the picro-formaldehyde mixtures from criticism:

"These are today the most widely used general purpose fixatives. They have nothing what-so-ever to recommend them save the demonstration of nuclei in meiosis, for which the original Bouin 1899, undoubtedly the best-known fixative at present employed, was developed."

Selective reading of Gabe's book is an excellent corrective for those who believe histological technique is cut and dried.

5.6.2   Which Fixative to Use?

Setting to one side Gabe's views on 10% formalin, his advice to try a small group of fixatives is very sound: including 10% buffered formalin in the group is recommended. For example, many commercial laboratories use Bouin's for testes and Davidson's for the eyes. Testes are particularly prone to shrinkage artefacts and many researchers choose to fix for 24 h in Bouin's before transferring to formalin. More recently, modified Davidson's fluid has replaced Bouin's. Modified Davidson's is also the choice for fixation of the eye as it reduces cellular shrinkage and improves resolution of the retina (J R Latendresse *et al., Toxicologic Pathology*, 2002, **30** (4), 524–533). Carleton provided advice on fixation on a tissue-by-tissue basis in the early editions of his well-known book. He limited his advice to the same groups of fixatives as chosen by Gabe with the addition of 10% formalin.

Given that there are so many variants of the basic mixtures we shall list them as:

1. 10% buffered formalin;
2. A picric acid–formalin–acetic acid mixture: Bouin's fixative as an example;
3. A mercuric chloride-containing mixture (Heidenhain's "Susa" if you like a complicated mixture, mercuric chloride–formalin if you prefer something less complicated but probably about as good. Carleton preferred "Susa" to every other micro-anatomical fixative and seldom listed it as anything other than first in his recommendations for various tissues. Drury and Wallington, editors of more recent editions of Carleton's book, found that mercuric chloride–formalin was about as good as the more complicated "Susa". "Susa", by the way, is a contraction of *Sublimat* (German for sublimate, *i.e.*, mercury) and *Säure* (German for acid). There was no Mr Susa, at least not in this context!
4. A dichromate–mercuric chloride mixture: Zenker or Zenker-for-maldehyde as an example
5. An alcohol–acetic acid mixture: Carnoy's fixative as an example

The formulae for these mixtures are given in the appendix.

These five fixatives include a broad range of primary fixatives and one would be unlucky indeed if at least one of the group did not produce very good fixation of any tissue chosen. Some histologists would be unhappy with this list and would assert that other, very good mixtures should have been included. For those wishing to try further mixtures, Gray provides a list of about five hundred.

A few points relating to the use of the five mixtures recommended here are provided in Table 5.4.

The real test of how well a fixative mixture fixes the tissues you are studying is to use it, to process the tissue to sections and to look at the results. An example of poor fixation *versus* good fixation is shown in Figure 5.1.

5.6.3 Deciding on the Quality of the Fixation

It is not possible to assess fairly the quality of fixation unless all the other steps in processing are both well carried out and held constant from fixative to fixative. Staining should be carried out in precisely the same way for all sections in the trial. Remember that you are trying to decide which works best on "your tissue".

Table 5.4 Standard fixative mixtures.

Fixative	Duration of fixation	Post-fixation washing	Notes
10% buffered formalin	24–48 h; can be longer without harm	None; go straight into 70% alcohol	Easy to use and not time-critical. Buffering is important: over-alkalinity is damaging. "Formalin pigment" may form in blood-filled tissue. The pigment can be removed with 1% ammonia in 90% alcohol: 1 mL "880 ammonia" per 100 mL 90% alcohol
Bouin	12 h or overnight	Wash in 70% alcohol, NOT in water	Very poor fixation of kidney, especially the cortex, but good for other tissues. Adding 1 g urea to 100 mL fixative is said to improve fixation of kidney tissue. Gurr, 1956, noted that fixation of mucin was poor. Distorts or destroys mito-chondria: not at all a cyto-logical fixative. Tissue can be left in Bouin for weeks without damage. The yellow picric acid staining can be removed from sections with 80% alcohol containing a few drops of sat-urated aqueous lithium carbonate
Zenker-formol	12 h or overnight	Wash in run-ning water overnight	Removal of mercury pigment from sections is essential: see Chapter 8
Carnoy	1 h at most; use thin blocks	Wash in sev-eral changes of 95% alco-hol, then de-hydrate in 100% alcohol	Perhaps this mixture could be deleted from the list on the grounds that it does not offer much that the others do not. It is, however, regarded as a cytological fixative. See section 5.9 for further cytological fixatives. Very effective for mitoses; damages mitochondria, as do all fixatives that contain acetic acid
Heidenhain's Susa	12 h	Transfer direct to 95% alcohol	Removal of mercury pigment from sections is essential. If material is transferred to lower grades of alcohol, swelling of collagen fibres is likely to occur. See Table 5.5 for the method for removing mercury pigment

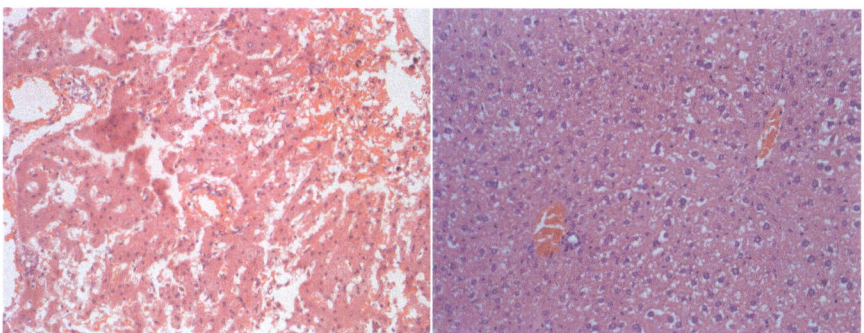

Figure 5.1 **The effect of poor fixation.** The left panel shows a poorly fixed piece of liver. Note the lack of structural characteristics and the presence of large amounts of blood within the section. Compare this to the well-fixed piece of tissue on the right, where the structure is preserved.

If you do not know what a section of a really well-fixed piece of tissue should look like, you have three options:

1. *Ask an expert to look at your sections and to decide for you which are best.*
   You will need to find an expert who is interested in the effects of different fixative mixtures: not so easy these days. If you do find an expert, ask him or her to look at the sections without knowing which fixative mixture was used for each section. This will produce a fair comparison and prevent plumping for the fixative used in his or her laboratory. Needless to say perhaps, you should ask for reasons as to why A is preferred over B. Asking your adviser to fix and process samples of your tissue can be helpful if you are really unsure of your technique.
2. *Compare your sections with pictures from publications of studies on the same tissue or with those in a textbook or atlas of histology.*
   The latter is not advised. Textbooks seldom say how the tissue used for the illustration was processed. Avoid atlases with colour drawings: these will tend to represent ideal appearances that may, in practice, seldom appear.
3. *Work from first principles yourself.*
   Of these methods this, the third, is particularly rewarding. Before setting out some ideas on how to do this, we shall consider the practicalities of fixation.

5.7   HOW TO FIX SAMPLES OF TISSUE

5.7.1   Immersion

Most tissue samples are fixed by immersion. The older books spoke of "throwing the tissue into fixative": placing it carefully in fixative is desirable. It is important for the tissue to be surrounded by a proportionally large volume of fixative: a 20 to 1 ratio is suitable. Tissue should not be allowed to rest on the bottom of a jar containing fixative: a pad of gauze will ensure that all surfaces are exposed to fixative. Gently agitating the bottle from time to time is sensible.

The golden rule for effective fixation is TAKE THIN BLOCKS. This is forced upon one if an automatic processing system is in use: the containers for the tissue will not accept blocks of more than about 5 mm in thickness. Cutting very large blocks (see Chapter 6) used to be a *tour de force* for histology technicians: sections of whole human brain perhaps. But, no matter how large the area of the block, if fixation is to be of high quality, the block must be thin.

5.7.1.1   Fixing the Lung.   Many histologists slice the fresh lung and place the slices in fixative. Inevitably, the lung collapses when removed from the chest and the sections will not show the lung in its normal state. For some purposes this does not seem to matter very much. But if detailed work on the histology of the gas exchange region is planned, the lung should be fixed in an expanded state. The difference between these two methods is shown in Figure 5.2.

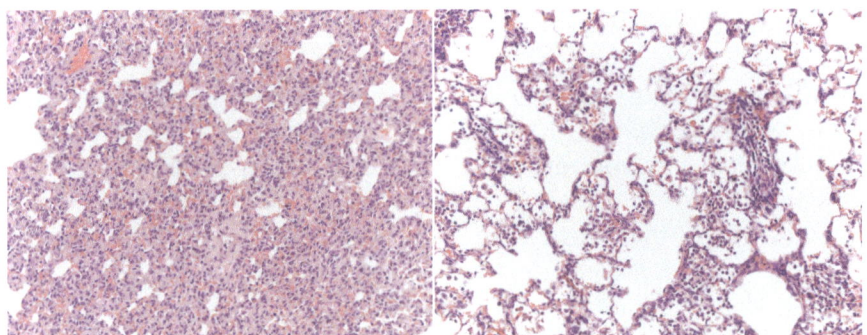

Figure 5.2   **Immersion *versus* inflation fixation of the lung.** The left panel shows the collapsed lung sample that usually results from standard immersion fixation of lung tissue. The right panel shows the bronchi and alveolar spaces, which are preserved by inflation of the lung tissue prior to immersion in fixative.

The simplest way to do this is to cannulate the trachea before the chest is opened at post-mortem examination and to infuse fixative under gentle pressure. This can be done using a large syringe or by means of a head of pressure provided by an aspirator placed about 45 cm above the bench. Fixing the lung whilst it is in the chest prevents over-distension. If the lung has been removed, expand it with fixative until the surface is smooth and the edges are sharp. The trachea should then be tied below the cannula and the cannula removed. The expanded lung should then be placed in a beaker of fixative and covered with gauze. The necessary duration of fixation will depend on the fixative used. One drawback of this method is that small amounts of oedema fluid are likely to be so diluted as to be difficult to see on microscopy. Recognising small amounts of oedema fluid is possible when sections of collapsed lung are examined but, ideally, the lung should be fixed in an expanded state but not filled with fluid. This can be done: both formaldehyde vapour and a mixture of formaldehyde vapour and steam have been used for fixation of the lung. These are rather specialised techniques and take some time to perfect. Fixation of the air-expanded lung *via* the pulmonary vasculature is probably the ideal method but, again, it is a complicated technique.

5.7.2 Fixation of Organs by Perfusion

Some tissues, such as the kidney and testis, undergo rapid changes after death: perfect fixation is thus difficult to achieve (Figure 5.3).

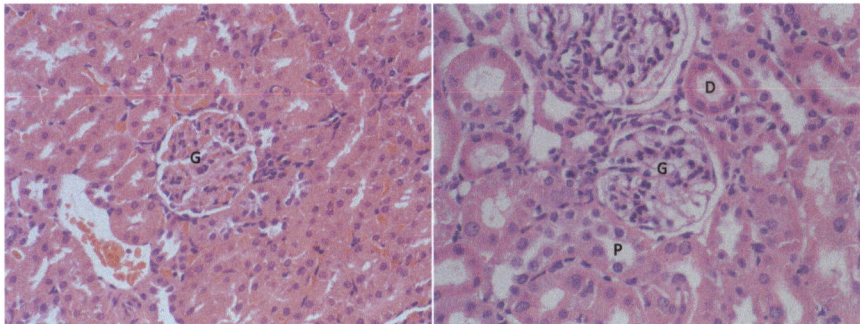

Figure 5.3 **Perfusion fixation *versus* immersion of the kidneys.** In the left panel is an immersion-fixed kidney; while the glomerulus (G) is distinguishable, the different types of tubules are not. The right panel shows that perfusion fixation can yield beautiful results as, in this example, the differences between the distal (D) and proximal (P) tubules can be clearly seen, as can the fine structure of the glomerulus (G). However, you do need to be aware of the difficulties of this technique and the artefacts that may result.

If large sections of the brain or liver are required and the organ cannot be cut into small pieces for fixation, then fixation by perfusion should be considered. Fixation by perfusion of fixative *via* the arterial system is rapid and effective. The animal should be deeply anaesthetised and the chest opened by a midline incision that runs from the abdomen across the costal margin. Ribs should be cut to allow wide access to the heart. Identify the left ventricle of the still-beating heart and introduce a cannula into it *via* the apex or anterior wall. This is tricky: the heart will be moving and speed is desirable. Holding the apex of the heart with a pair of forceps is helpful but be careful: the heart of a small animal, such as a mouse, tears rather easily. The plastic cannula should be cut at an angle, like the needle for a hypodermic syringe, but should not be too sharp. The cannula should be advanced into the root of the aorta. Tying the cannula in place is helpful but not essential. A curved suture-carrier should be manipulated around the root of the aorta and the suture tied around the cannula. Effective perfusion can be achieved in a mouse by introducing a large calibre needle into the left ventricle.

The cannula should be connected *via* a three-way tap to a reservoir of 0.9% saline, preferably at 37 °C (or slightly above to allow for cooling in the cannula), and to a reservoir of fixative. The reservoirs should be about 30 cm above the animal. Before introducing the cannula into the heart, saline should be run through to ensure that no air will be introduced into the vasculature: air bubbles block small vessels. Saline should be flowing slowly through the cannula as it is introduced into the heart. Once the cannula is in place, snip the right atrium with fine pointed scissors, open the three-way tap to the saline and perfuse the animal. Fluid will run from the right atrium. Continue until the fluid running from the right atrium is clear of blood and the liver is pale. Now switch to the fixative and run through, not too quickly, about 100 mL. This is enough for a rat; rather less would be enough to fix a mouse. Contraction of skeletal muscles will be observed and the liver will become firm. Remove the cannula and place the specimen in a bath of fixative for 12 h. Alternatively, remove the organs required for study when the perfusion is complete.

Heparin and vasodilator substances are sometimes included in the saline perfusate: this does not seem to be necessary.

Fixation by perfusion is the best method for fixing the brain. It also makes removal of the brain very much easier because it becomes firm and easy to handle.

5.7.3 Fixation of Tissues that Curl or Contract

Fresh skeletal muscle will contract if placed in fixative. Allowing the muscle to rest under a piece of saline-soaked gauze for 30 minutes before being placed in the fixative is helpful. If the muscle has a tendon at each end then it can be tied to a cocktail stick or across the prongs of a U-shaped frame before being placed in the fixative: this will maintain the muscle at its normal length.

Pieces of gut roll up in fixative and pinning out the gut wall is necessary if first-class sections are to be prepared. Pieces of gut should be opened longitudinally, washed gently with saline, pinned to a cork slip with the serosal surface towards the cork and placed in fixative. The cork will float: ensure it floats with the tissue downwards and in contact with the fixative and immerse it by covering with gauze. Ordinary pins are satisfactory unless a fixative containing mercuric chloride is used: mercuric chloride corrodes the metal of the pins. For those who like classical methods, pins should be replaced with hedgehog quills or thorns from hawthorn trees. Hedgehog quills may be, rather obviously, obtained from a hedgehog that has been killed on the road. Snip off a large number of quills, wash in dilute phenol and then in water, dry and store in a jar. Quills can be used again and again. (The quills can also be used as needles for mechanical gramophones and, though they wear out rather quickly, do less damage than steel needles to aged and valuable re-cords: a fact seldom encountered in books on histological technique.) For those who are rather more up to date: plastic pins are perfectly satisfactory.

Another option for getting the length of the gut onto a single slide is called a Swiss roll. As shown in Figure 5.4, this allows the researcher to see along a large portion of the length of the gut in a continuous piece. This is done very basically by gripping one end of the gut between two cocktail sticks and rolling it up on top of itself. Once you have reached the desired length, a pin is stuck through the roll and the sticks removed. The whole roll is then put back into fixative for processing to wax.

After preliminary fixation, a common method for handling tissues is to trim them into smaller pieces and contain them within histo-logical cassettes so that the tissue is not over-handled (which may introduce artefacts). A well-organised tissue trimming station (Figure 5.5) can be of great help if you are processing several different tissues, especially if there is tissue that needs special handling. If the tissue is going to shrink, or does not fill the cassette, like thin slices of

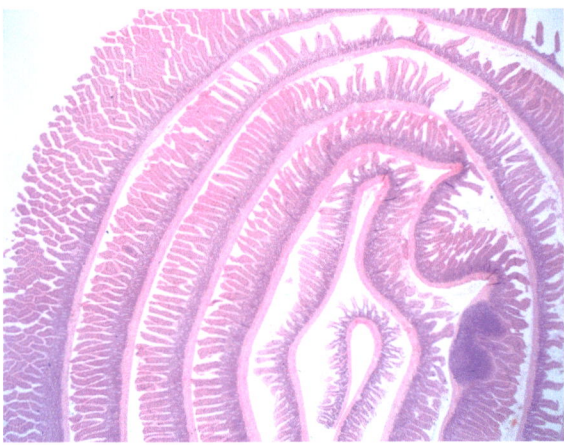

Figure 5.4 **Swiss roll.** If you like your intestines in large amounts, this technique allows you to get nearly a whole rodent small intestine onto one slide.

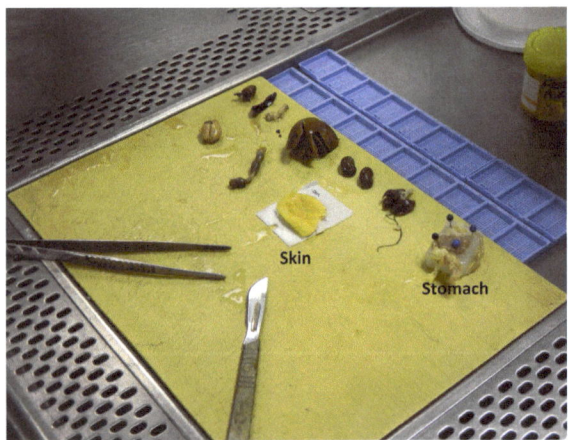

Figure 5.5 **Tissues fixed and ready for trimming.** Note how the stomach has been pinned out on a wax block and the skin sample stapled to a piece of card to prevent curling during the fixation process. You can easily make you own wax blocks out of the mounting wax.

brain or embryonic tissues, pieces of sponge can be inserted into the cassette on top of the tissue to hold it in place. (A note regarding the sponge—any sponge will do as long as it is not too thick for your cassette, or too dense to allow penetration of the fixative and wax.)

5.8 EXAMINING SECTIONS TO ASSESS THE QUALITY OF FIXATION OF THE TISSUE

All fixation changes the appearance of tissue. Some changes, for example, coagulation of protein, are desirable and essential to good fixation. Others, for example, dissolution or glycogen or lipids, are inevitable with some fixatives but can be avoided by selecting fixatives carefully. Wallington's review of artefacts in paraffin wax sections provides much information on this subject (E A Wallington, Artifacts in tissue sections, *Medical Laboratory Science,* 1979, **36**, 3–61).

5.8.1 Diffusion Artefacts

Fixative diffuses through tissue and may carry with it soluble materials. Glycogen is easily displaced in this way and may come to lie against the sides of the cells "away from" the direction of diffusion of the fixative. This can be very obvious. The effect is seen with fixative containing alcohol (a non-additive fixative) to a greater extent than with fixatives containing additive fixatives, such as formaldehyde. Diffusion artefacts may also be seen in the colloid of thyroid follicles. Markedly different staining reactions occur in the colloid due to movement of the protein of the colloid. The colloid thus varies in terms of its protein content in the direction of diffusion of the fixative: more protein is found at the side of the follicle away from the side of entry of the fixative (Mayer, see Bibliography). Coagulation of the colloid leads to its pulling away from the surfaces of the cuboidal epithelial cells: an artefact known as scalloping. Colloid is difficult to fix: fracturing occurs. The fracture lines lie parallel to the edge of the knife used for cutting the sections. If the thyroid is being studied then selecting a fixative that minimises these artefacts is obviously sensible.

5.8.2 Distortion and Shrinkage Artefacts

Shrinkage is almost inevitable in paraffin wax sections. Distortion of structures is also common-place but can be reduced by using appropriate fixatives. Formalin is much less effective than the mercuric chloride or dichromate fixatives as a preventer of distortion. A sign of shrinkage is the appearance of spaces in the tissue. These are seen around blood vessels: the cells of the tissue pull away from the vessels. Such spaces are often seen in sections of poorly fixed lung: shrinkage leads to the formation of spaces around branches of the pulmonary arteries and bronchi or bronchioles. The loose connective tissue surrounding these structures contains lymphatic vessels, which become distended during the early stages of development of pulmonary oedema. Distinguishing the lymphatic vessels from artefactual spaces or cracks in the loose connective tissue can be difficult. Of course, the lymphatic vessels are lined by endothelium and the artefactual spaces or cracks are not, but seeing the cells of the endothelium is not always easy. Similar pulling away of tissue from blood vessels may be seen in sections of poorly fixed liver. A gap appears between the hepatocytes and the very thin and incomplete walls of the hepatic sinusoids. This gap is very small, indeed hardly visible, in well-fixed material. It will be known that this space always exists: it is the space of Disse. However, although it is visible on electron microscopy, it is too small to see with light microscopy. In well-fixed material the space cannot be seen with the light microscope.

5.8.3   Pigment Formation

Several fixatives cause pigment to be deposited in tissue. The pigment is, of course, an artefact, but it is not a sign of poor fixation: it is simply an inevitable consequence of using certain fixatives. Distinguishing this artefactual pigment from endogenous pigment, such as melanin or haemosiderin, is important. Distinguishing artefactual pigment from particles of coal dust in the lung or in the macrophages of the spleen is also important. Artefactual pigment lies **on** rather than **in** cells: this can often be detected by focusing up and down. Pigment appearing as a result of fixation can be removed. Table 5.5 sets out how this can be done.

   In addition to looking for artefacts, the staining of nuclei, cytoplasm and cytoplasmic organelles should be examined. The chromatin of nuclei should be brightly and sharply stained. Collagen fibres should also be brightly stained by trichrome methods and should appear wavy and fibrillar in structure. Fat bundles of rather amorphous collagen are a sign of poor fixation. As with all histological techniques, the ability to recognise well or poorly fixed material develops with practice. Every section examined should be looked at with an eye to the quality of fixation.

5.9   CYTOLOGICAL FIXATIVES

The great days of cytology using the light microscope are perhaps over: the electron microscope has revolutionised the study of cell

Table 5.5 Removal of pigment artefacts.

Fixative	Pigment	Removal
Formalin	Black pigment formed as result of combination of formalin with haematin. Found in organs containing a lot of blood, *e.g.*, the spleen or bone marrow	Treat sections with alcoholic picric acid for 20 minutes, wash well in water. 1% ammonia in 90% alcohol also works well
Mercuric chloride	Mercurous chloride	Treat section with 0.5% iodine in 70% alcohol for 2 minutes. Then place in sodium thiosulphate (7.5 g sodium thiosulphate in a mixture of 900 mL distilled water and 100 mL 96% alcohol). Rinse in water
Dichromate	Black pigment, composition unknown	Treat sections with 1% HCl in 70% alcohol for 30 minutes. Wash in water
Picric acid	Pigment produced by basic aniline dyes depositing insoluble material is sections	This pigment will not be present unless basic aniline dyes are used. If they are used and the pigment appears, go back to unstained sections and wash really well in 70% alcohol to remove all the picric acid. Do not wash with water as this may remove the proteins that, despite being bound to picric acid, remain soluble in water. 80% alcohol containing a few drops of saturated aqueous lithium carbonate is effective

structure and debates about the Golgi apparatus and the reality or otherwise of intracellular canaliculi are now of only historical interest. It is interesting, however, to note that osmium tetroxide, the fixative used by the early cytologists, is still much used by electron microscopists, especially after fixation with glutaraldehyde.

Fixatives like Bouin and Carnoy provide very good fixation of nuclear material and may be used for the study of chromosomes and of mitotic figures. Sanfelice's fixative, which is not quite one hundred years old, is also excellent (the formula can be found in the Appendix). Osmium tetroxide is an unpleasant material: it gives off an irritant vapour and is poisonous. It should be handled with care, gloves and in a fume cupboard. It is, however, an excellent cytological fixative and the formula for Flemming's fixative mixture is included in the Appendix. Compared with other fixatives, it has significant drawbacks: small pieces of tissue must be used, the outer parts of the tissue tend to be over-fixed, the tissue is blackened and some stains, for example, alum haematoxylin, do not work as well as they do after other fixatives. These are not inconsiderable drawbacks. But fixation at some distance from the surface of the tissue will be perfect and both nuclei and cytoplasmic structures will be superbly demonstrated by Heidenhain's iron haematoxylin stain. For this reason, it has been included here, though it is unlikely that more than a few readers will use it. Small pieces of tissue should be fixed in Flemming's fixative for 24 h and then washed well in tap water before being dehydrated and processed for paraffin wax sections.

5.10 TISSUE PROCESSING

Once the tissue is fixed, it would theoretically be firm enough to be cut into a sufficiently thin section to be useable under a microscope, but in order to do that we need to be able to mount the specimen so that it can be handled by our chosen method of mechanical sectioning. Paraffin wax is the medium most widely used for infiltrating and enclosing tissue prior to sectioning, although other media, like celloidin, resin, and plastics (methacrylate), are used on occasions.

  Paraffin wax has been in use for over one hundred years and is still in use in all pathology laboratories. Routine sectioning for toxicological histopathology is always done using paraffin wax. Of course, the technique has drawbacks: tissue has to be fixed, dehydrated and cleared before it can be infiltrated with wax. This rather demanding sequence leads to tissue shrinkage and, in some cases, to distortion. Exposure to molten paraffin wax at about 56 °C leads to further shrinkage. In addition, paraffin wax is a fairly soft inclusion medium. It is very suitable for tissues that, after processing, are about as hard as the wax, but for harder tissues a tougher medium is needed. One of the great advantages of paraffin wax is that ribbons of serial sections can easily be produced. This advantage should always be remembered and will be obvious if, after cutting paraffin wax sections, the reader turns to frozen sectioning or to celloidin as an inclusion medium. Five micron sections can be cut without difficulty using paraffin wax. This is a very convenient thickness for light microscopy: nuclei are about this size and, even from a tissue containing many closely packed cells, not too much superimposition of nuclei will occur, thus examination with the light microscope is expedited. Thicker sections are sometimes needed, for example, in the case of lung tissue, and sections of up to about 20 μm in thickness can be prepared. If sections of much less than 5 μm are needed then paraffin wax ceases to be the ideal medium: plastic sections take over. For electron microscopy, very much thinner sections, 50–500 nm, are needed. To prepare these, very tough inclusion media are required: resins, for example. Paraffin wax is a very useful medium to work with when learning histological techniques: sectioning is easier than in the case of frozen material and the formation of ribbons of sections makes the process much faster than cutting celloidin sections. Most histologists start with paraffin wax and then move on to other techniques if necessary.

The great drawback of paraffin wax is that it is not miscible with water. Before our fixed tissue can be impregnated or infiltrated with wax, all the water in the tissue must be replaced with a solvent that is miscible with wax. The solvents that are suitable for this, for example, xylene, toluene or benzene (and many others: see section 5.10.1), are unfortunately not miscible with water. Fortunately for histologists, water is miscible with alcohol, and alcohol is miscible with solvents that are miscible with wax. Thus, a step-by-step approach is adopted: water to alcohol to solvent to wax.

All commercial laboratories are equipped with a battery of tissue processors that can be programmed to deal with the requirements for the processing of tissues from all creatures: great and small.

Unfortunately for the occasional histologist, these processors (Figure 5.6) cost tens of thousands of pounds, so you will probably need to do this by hand. A step-by-step protocol for processing is described in the following sections. The first step of the process is dehydration with alcohol.

5.10.1   Dehydration with Alcohol

Well-fixed tissue should be dehydrated using what are usually referred to as ascending grades of ethyl alcohol. This is such a standard routine that one might wonder whether it has ever been questioned. Masson regarded the use of ascending grades as pointless and transferred tissue after fixation and whatever washing was specified into absolute alcohol; Gabe followed Masson's lead. Carleton noted this but argued that, whilst the Masson approach might be satisfactory for many tissues, it was unlikely to be as satisfactory for delicate tissues as the usual, *via* the grades approach. He argued that shrinkage was likely to be more marked if tissue was transferred from water to absolute alcohol than from water to 50 or 70% alcohol and thence up the grades to absolute. Drury and Wallington specified the use of ascending grades in their schedules for hand and automatic

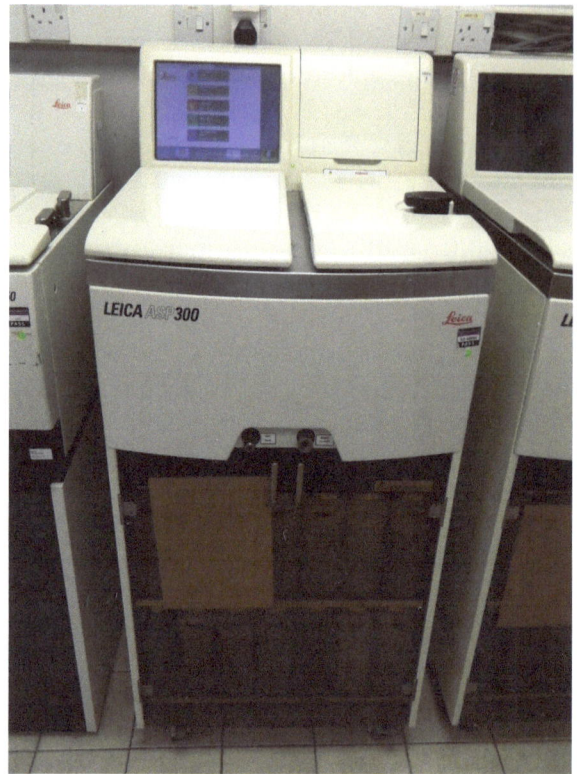

Figure 5.6   **Standard laboratory tissue processor.**

Table 5.6   Dilution of alcohol. Modified from Carleton and Drury, 1957.

Grade required (%)	Amount of 95% alcohol (mL)	Amount of distilled water (mL)
90	940	60
80	836	164
70	730	270
50	520	480

processing. It is recommended that the standard, up the grades approach is used.

Absolute alcohol is expensive and it is usual to make up the 50, 70, 80 and 90% alcohol from industrial methylated spirit, which contains about 95% ethanol. In this book 95% alcohol means industrial methylated spirit.

Table 5.6 summarises how to make up 1 L of whatever grade of alcohol is needed.

If tissue has been fixed in Carnoy's fixative or in Heidenhain's Susa, it may be transferred to 95% alcohol without passage through the lower grades. The following schedule recommended by Drury and Wallington is perfectly satisfactory.

Fixation
Wash in water if necessary (see details of fixatives)
70% alcohol   8 h (during the day)
90% alcohol   Overnight
Absolute alcohol   2 h
Absolute alcohol   2 h
Absolute alcohol   2 h
Then transfer to the clearing agent: see section 5.10.2.

Note that there is no need to begin with 50% alcohol unless very delicate tissues are being processed. Different authorities give very different advice on the time needed in each of the grades of alcohol. In 1957 Carleton recommended 24 h in each grade; if blocks are small (3 mm or so in thickness), the times given above might be halved. The overnight period in 90% alcohol is not much more than a convenience in that it follows a working day in the 70% alcohol and avoids the need for changing fluids during the evening and night. Agitating the jars containing the tissue and the alcohols speeds the process. Schedules for use with automatic processors (an example of which is shown in Figure 5.6) are set out in Drury and Wallington (see Bibliography).

If you are not in a hurry, moving the tissue from one grade of alcohol to another at the start and end of the working day will produce good results.

5.10.2   Clearing

The word clearing refers to the fact that some "clearing agents" render tissues clear. But the clarity of the tissue has nothing to do with the real objective, which is to replace the alcohol with a solvent that is miscible with paraffin wax. Though there is not much to argue about with respect to the dehydration sequence (different alcohols and some other substances can be used but do not seem to add much and are probably not worth the trouble), there are dozens of clearing agents—each with its own advocates. Clearing is achieved by transferring the tissue from absolute alcohol to the first of several baths of clearing agent. Because the clearing agent is not miscible with water, it is important to ensure that dehydration is complete. Some clearing agents are "tolerant" of a little water in the tissue, others are not. If the clearing agent becomes cloudy this suggests that dehydration has not been satisfactorily complete. The following factors have determined the preferences of experts for various clearing agents:

1. How much the clearing agent hardens the tissue;
2. How quickly the clearing agent acts;
3. Whether the clearing agent is "tolerant" of a little water in the tissue;
4. Whether tissue can be left in the clearing agent for long periods without becoming ruined by excessive hardening;
5. Whether it is essential to remove every last trace of the clearing agent during infiltration of the tissue with paraffin wax;
6. Whether the clearing agent imparts an indefinable quality to the paraffin wax block that leads to better sectioning;
7. Toxicity;
8. Cost.

Two broad groups of clearing agents can be identified:

1. The organic solvents, such as xylene, toluene, benzene, petroleum ether, chloroform and carbon tetrachloride;
2. The "essential oils", such as the oils of cedar, bergamot, sandal wood, cloves, origanum, cajeput and thyme. To this rather botanical group we might add beech-wood creosote, methyl salicylate (oil of Wintergreen) and terpineol (derived from terpenes found in plants). These are, of course, organic compounds and solvents but, for convenience, have been separated from the organic solvents listed above.

Some readers will wonder about these essential oils and ask whether they are still in use. In general those listed above are not used today, but much used commercial clearing agents, such as Histo-Clear, contain oils distilled from oranges (the essential oil is limonene). This, unsurprisingly, explains the pleasant, if sometimes rather strong, orange-grove smell of the agent and often of the laboratory where it is in use. The essential oils have the great advantage that they are regarded as non-toxic, at least in comparison with xylene, benzene, chloroform and carbon tetrachloride. The oils are also "tolerant" of a little water being left in the tissue at the start of clearing. This is not the case with xylene, toluene or benzene: with these clearing agents, care should be taken to ensure complete dehydration of the tissue. Carbon disulphide is an effective clearing agent but its inflammability and smell make it little used. Lillie (1976) used gasoline (aviation spirit that is essentially petroleum ether) with "great satisfaction". He found that very little hardening of the tissue was produced. The high inflammability of petroleum ether needs to be borne in mind.

Once the tissue is cleared, it can be infiltrated with molten paraffin wax. This is done by placing the tissue in molten wax in an oven or, if automatic processing is used, by transfer of the tissue to a heated container of wax. The organic solvents are easier to remove from the tissue during immersion in molten paraffin wax than are the essential oils. The solvents evaporate from the wax baths, while the oils have to be washed out of the tissue by three or four changes of wax. This makes the process longer when oils are used. In addition, the wax becomes contaminated with the oils to a greater extent than with the solvents and the wax baths have to be changed rather frequently. Evaporation of organic solvents from the wax baths can lead to exposure of staff to the vapour: histology laboratories used to smell of benzene when tissue that had been cleared in benzene was being infiltrated with wax in an oven placed on the open bench. This is most undesirable: the oven should be placed in a fume cupboard. Speeding up clearing by putting beakers of benzene containing large blocks of dehydrated tissue in wax ovens on the open bench would not be allowed today. Some workers recommend transferring the tissue from the clearing agent to a mixture of clearing agent and molten wax and from that to pure wax. This is not generally necessary, though it can be argued that it reduced shrinkage.

Of the many clearing agents that have been recommended by various authorities only three will be recommended here: toluene, chloroform and cedar oil. Xylene should not be used because, although it penetrates quickly and is easily removed, it hardens tissue undesirably. Chloroform and toluene produce less hardening, though they penetrate the tissue less rapidly than xylene. Chloroform leads to little shrinkage of tissue in the paraffin baths and chloroform has an advantage over toluene in that it takes up water (due to better "water tolerance"), thus less complete dehydration is required. Chloroform or toluene will produce good results with most tissues.

The schedule recommended by Drury and Wallington for processing tissue by hand, and included in part above, can now be completed:

Fixation
Wash in water if necessary
70% alcohol   8 h (during the day)
90% alcohol   Overnight
Absolute alcohol   2 h
Absolute alcohol   2 h
Absolute alcohol   2 h
Toluene or chloroform   Overnight, 16 h
Three changes of molten paraffin wax
Embed in fresh wax
For schedules for automatic processing: see Drury and Wallington.

5.10.2.1   The Use of Cedar Oil.   Cedar oil, better referred to as cedar-wood oil to avoid confusion with the preservative oil, which people use on their cedar-wood garden furniture, is expensive: a search undertaken in August 2012 revealed that a litre of cedar-wood oil costs about $300. This makes it too expensive for most users of automatic processors. But for processing small amounts of tissue for research work, rather than routine examination, it has advantages: it clears from 95% alcohol (meaning that complete dehydration is not absolutely necessary, though tissue should always be transferred from 95% alcohol to absolute alcohol before being cleared), complete removal in the wax baths is not necessary (blocks can smell slightly of cedar oil), the sectioning properties of the wax blocks are improved by there being a little (but only a little) cedar oil present in the wax, minimal shrinkage occurs in the paraffin baths and tissue can be left in the cedar oil for weeks or months without ruining it. It is also very good for "difficult tissues", such as the uterus, human skin and tendon. Transfer of the tissue from the cedar oil to benzene or xylene, for 30 minutes, before proceeding to the first wax bath speeds up elimination of the cedar oil.

The cedar oil that is used for clearing is thinner than the oil used for oil immersion microscopy. Several authorities point out that the oil intended for immersion microscopy should NOT be used for clearing tissue. Bolles Lee, in the fourth edition of his book *The Microtomist's Vade-Mecum*, took quite the opposite view: "I always use the *thickened* oil as supplied for use with immersion objectives"

(Bolles Lee's italics). This advice was dropped from later editions (edited by others). The thinner oil works well today: perhaps the quality has improved since 1896! A schedule for use of cedar-wood oil is given below.

Hand processing using cedar-wood oil
    Dehydrate
    Cedar wood oil 1   Overnight
    Cedar wood oil 2   Overnight

If small pieces of tissue are being processed then large volumes of cedar wood oil are not required: 50 mL is enough to clear a few small blocks.

It is unlikely that many readers will go to the trouble of obtaining and using cedar-wood oil. The commercial clearing agents are probably as good, but if difficulties occur it might be useful to go back and try a range of clearing agents. Under these circumstances it would be sensible to try cedar-wood oil.

The final stage of processing is to infiltrate your well-fixed, dehydrated and cleared tissue with molten paraffin wax. The wax permeates the tissue and replaces all other fluids. Thus, the clearing agent is driven out and replaced by wax. The wax enters all parts of the tissue: it enters the cells and their nuclei. Substances like protein remain, of course, but they are intimately surrounded and permeated by the wax. Lipids will, of course, have disappeared during dehydration: the spaces they leave are filled with wax.

5.10.3   Infiltration or Impregnation with Paraffin Wax

5.10.3.1 The wax. Paraffin wax is a by-product of oil refining. The wax used for histological work in countries like the UK melts at about 56 °C; in a warmer climate a wax with a higher melting point should be chosen. Some wax that melts at about 45 °C should also be available in the laboratory. Low-temperature wax can be useful for some immunohistochemistry applications where the epitope is heat labile. Wax may be bought as pellets: these are convenient and melt rapidly in the wax oven. The oven should be set at just above the melting point of the wax; 60 °C is a suitable temperature for ordinary work. Most books recommend filtering the molten wax before use. This can be done in the oven using a wide-stemmed glass funnel and coarse filter paper folded so that it forms a fluted cone. It is important that there should be no grit or debris in the wax. Wax with an added plasticiser (for example, Paraplast®) is particularly useful when you get on to cutting your sections as it is easier to ribbon (see Chapter 6) than ordinary paraffin wax.

There are automatic tissue processors available that provide baths of molten wax into which the tissues, in cassettes, can be transferred from the clearing agent. If a by-hand approach is being used then tissue should be soaked for one hour in each of three beakers of wax in the wax oven. It is a mistake to leave the tissue in the molten wax for longer than necessary: hardening and shrinkage occur and the blocks become more difficult to cut. If, at least at first, the reader uses small blocks, not more than about 5 mm in thickness, then an hour in each of the three wax baths will be satisfactory. Wax will take up the clearing agent and must be discarded after a few uses. Schedules for automatic processing will be found in the manuals of technique: see Bibliography.

5.10.3.2 Vacuum Embedding. Vacuum embedding is a refinement whereby this part of the process is performed under reduced pressure, which allows the infiltration process to be shortened. This process is always known as vacuum embedding, but vacuum infiltration or vacuum impregnation would be better phrases.

This is easily accomplished using a small vacuum impregnation oven. The pressure within the oven is reduced by a filtration pump attached to a cold water tap *via* a thick-walled flask to prevent the possibility of water flowing into the oven. Vacuum embedding is very effective if the tissue contains air: lung is the only example. It is also useful when tissue that is penetrated only slowly by wax is being infiltrated. Brain and tissue containing much collagen, for example, tendon or fibrous tumours, or muscle can all be infiltrated much more quickly under vacuum. Reducing the time spent in the wax is not only convenient; it also reduces the amount of hardening of the tissue that occurs. Not leaving tissue too long in molten wax is sensible, even if some experts advise this makes little difference to the blocks.

A few things to remember:

- Always reduce the pressure within the oven slowly at the start of infiltration and always raise the pressure within the oven slowly at the end of infiltration because too rapid changes in pressure will damage delicate tissue, such as lung;
- Letting water flow back from the filter pump into the embedding chamber would be disastrous: always include a thick-walled flask between the pump and the chamber.

Drury and Wallington recommend a pressure of 40–65 kPa (about half an atmosphere, or 380 mmHg) in the chamber. Measure the pressure with a mercury manometer and close the tap connecting the chamber with the pump as soon as this pressure is reached. You do not need to run the pump for the whole period of embedding.

Your tissue has now been processed, you can now move on to blocks and then sections.

Paraffin Wax: Embedding and Section Cutting

In Chapter 5 we looked at preparing samples of tissue for embedding in paraffin wax. Here, the process of embedding and sectioning of the wax blocks, attaching to a slide and covering with a coverslip is considered. It is often said that cutting high-quality sections is an art that can be learnt only from a master and only after long practice. We agree, but think that anyone who has a genuine interest in histology can and should learn how to cut sections of acceptable, if not remarkable, quality. This is a challenge for the beginner. As in all laboratory methods that involve a number of stages, it is essential to take great care at each stage. Damage done early on in the process can often not be rectified later on. Cutting the sections is described in this chapter. We have chosen to describe in detail the technique for using just one type of microtome and have chosen a type of instrument in wide use today: the rotary microtome. Practice is needed before first-class sections can be prepared on a regular basis from a range of tissues. The beginner should start with an "easy" tissue and work up to more demanding tests.

Handling frozen sections is also considered, albeit briefly.

Histological Techniques: An Introduction for Beginners in Toxicology
By Robert Maynard, Noel Downes and Brenda Finney
Published by the Royal Society of Chemistry, www.rsc.org

6.1   PARAFFIN WAX: EMBEDDING AND SECTION CUTTING

Once completely infiltrated, it only remains to mount the tissue in a manner that is suitable for sectioning on a microtome. The standard way to do this is to produce a wax block. The processed tissue is placed in a container of molten wax and the wax allowed to solidify. This produces the block, which can then be cut into very thin sections.

6.2   EMBEDDING

Once the tissue is fully infiltrated with wax, it is ready for embedding. This means casting a block of wax with the tissue inside it.

If a Tissue-Tek® embedding centre (or something similar from other manufacturers) is available then embedding the tissue in wax blocks is easy (Figure 6.1).

Standard metal container molds are used, a cold plate is available for chilling the container of the tissue and the molten wax, the molten wax is dispensed *via* a tap and heated forceps are provided for orientating the tissue in the wax. One could hardly ask for more. Remember to smear the inside of the metal container mould with glycerin or liquid paraffin: this makes easier the removal of the block (Figure 6.2). All laboratories handling a lot of material use such a system.

It is possible that the research worker might not have access to such apparatus; in that case the following method will produce perfectly acceptable results, if rather more slowly.

Embedding can be done in any small box, from a matchbox to a paper box made by elementary origami: see section 6.4.4.1. The box should be made of paper or cardboard: it will be peeled away from the wax block once the block has set. If you search around some of the

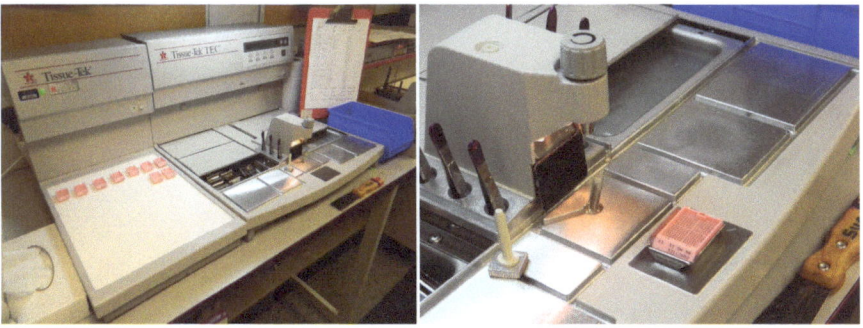

Figure 6.1   An embedding centre.

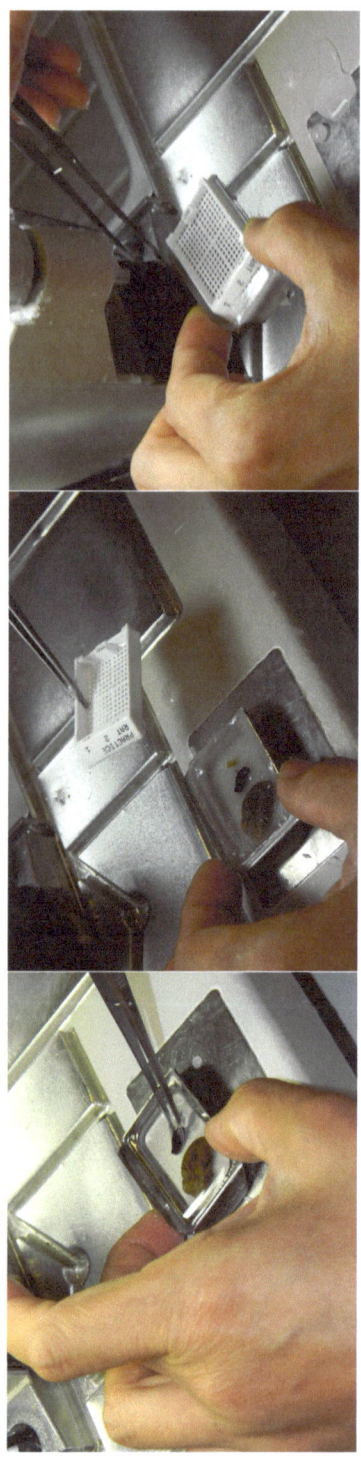

Figure 6.2 **Using an embedding centre.** Align the tissue in the hot wax and hold against the bottom until they stick (left). Top up with hot wax and get your cassette (middle). Place cassette into warm wax (right); it sets quite quickly so you will need to be fairly swift.

older publications you may find a reference to Leuckhart moulds. These comprise two L-shaped pieces of brass or any other metal that can be arranged on a piece of glass or metal to produce a cavity of suitable size. Several pieces of tissue can be embedded in the mould and cut out later.

The first step is to fill the mould with molten wax; smearing the mould and the glass or metal base-plate with glycerin before pouring in the wax is a good idea, but not essential.

Remove the tissue from the wax bath (in the 60 °C oven) with warm forceps. Warm the forceps in a Bunsen flame: do NOT make them red hot as they will then damage the tissue, just warm enough so the wax does not solidify on them immediately.

Lower the tissue into the wax in the mould so that it lies against the glass plate. It will actually be separated from the glass plate by a thin layer of wax, which will be solidifying. The side of the tissue from which you wish to take sections should be placed downwards, towards the glass or metal plate. Work quickly: the wax will be setting.

Orientate the tissue. If the tissue block is square then bring one edge parallel with a side of the mould. There is much to be said for cutting triangular blocks of tissue and for orientating them so that a point of the triangle is towards one edge of the mould. When the tissue is cut, the knife will strike the point of the block first, which makes cutting easier.

Attach a label by pressing a slip of paper, with the necessary details written on it in soft pencil, into the edge of the mould.

Now pick up the plate and blow gently on the surface of the wax. A skin will form.

Tilt the mould and lower it, at a shallow angle, into a large volume of cold tap water. A hand-basin or sink of cold water is suitable. Do not use iced water: this can cause cracking of the wax block. Leave the mould in the water for a couple of hours. Immersion of the angled mould (the mould and its base should be "sunk" like a ship sinks at sea: at a shallow angle) will prevent water entering the wax and the molten wax escaping by floating up in the water. If this occurs, the final wax block will have a hole in its surface: very undesirable. Leave the block in the water for a few hours, over lunch perhaps.

Remove from the water, tap on the bench and remove the wax block.

6.2.1 Trimming the Wax Block (not Needed if an Embedding Centre and Plastic or Metal Embedding Moulds have been Used)

It is a good idea for the beginner to start with one piece of tissue in a fairly small wax block. Place the block on the bench, preferably on a board, so that the surface from which you will be cutting sections is uppermost. You will be able to see the tissue through the thin layer of wax that was at the bottom of the mould. Take a one-sided safety razor blade and pare away the wax from around the tissue so as to create a square surface with the tissue in its centre. Take care: pare away the wax slowly. Two things are important:

1. There should not be more than 1 mm or so of wax around the tissue. To begin with, I would leave a bit more than this: say 2 mm.
2. Think how you will be cutting the block: decide on a top and a bottom and make these surfaces strictly parallel with one another. If you don't do this, the ribbon (see section 6.3) will be curved. Some workers nick off one of the top edges of the block so as to be able to identify this surface.

If you have put a number of pieces of tissue into the mould, you will need to decide whether you want to cut these all together or to prepare individual blocks for sectioning. Your size limit is set by your expertise in sectioning (big blocks are more difficult than small blocks) and the size of slide you plan to use. If you decide to cut out a number of blocks, then cut up the main wax block with the single-sided safety razor blade. Be careful: do not fracture the block.

6.2.2   Fixing the Block to the Chuck of the Microtome

If you are using an automated system, your blocks will be held on a plastic support that slots neatly into the special holder attached to the chuck of the microtome. For those less fortunate, who by now will be becoming envious, the wax block needs to be attached to the chuck. This is done by fusing the wax of the block with a layer of wax placed on the surface of a block-carrier, which fits into the chuck. Metal block-carriers are standard; as may be divined from Gray's comments regarding the metal block-holder supplied with the Spencer rotary microtome, this is far from ideal.

"Since the majority of sections today are cut on a Spencer rotary microtome, we will describe the use of one of the holders supplied with the machine, though the ingenuity of man has not yet succeeded in devising a worse way of attaching a paraffin block to a microtome."

The problem, and it is a problem with metal block-carriers in general, is that it is difficult to persuade the wax to stick to the metal. Wooden blocks can be cut so as to fit into the horizontal jaws of the chuck of a rotary microtome. These wooden blocks should be cubes of about 1 inch with the front face, the end grain of the wood, being marked horizontally and vertically with shallow saw cuts 1/8 inch apart. These work rather well in a low-tech way.

Dip the wooden block or metal block-carrier into molten wax and build up a layer of wax on its face. Take a wooden handled spatula and heat it in the Bunsen flame until hot enough to melt the wax easily, but again not red hot. With the block-carrier held in a vice (the wooden blocks sit nicely on the bench but the metal carriers often do not), take the spatula in your right hand and the wax block in your left. Place the blade of the spatula against the face of the block-carrier and bring down onto it the wax block. Ensure that the edges of the wax block are parallel with the edges of the wooden block. In the case of metal carriers this is not necessary because they can be rotated in the chuck of the microtome. Slip out the spatula blade and press the wax block firmly to the carrier. Hold for a few moments then immerse in cold water for 30 minutes. Wood plus wax floats so you will need to hold it under the surface with a pad of cotton wool.

6.2.3   Double Embedding

Double embedding is an excellent method that is little used today. However, for sectioning difficult tissue it has no equal as it combines the support provided by celloidin with the easy cutting of ribbons of sections characteristic of wax embedding.

Method (for a small block of tissue)

Make up 100 mL of a 1% solution of celloidin in methyl benzoate. The celloidin is bought as a white powder damped with butyl alcohol. The dampening alcohol is ignored in making up the solutions. Celloidin is a form of cellulose nitrate and is explosive when dry: hence the damping with alcohol. It will take a few days for the celloidin to dissolve in the methyl benzoate: keep the bottle or flask stoppered.

Soak dehydrated tissue in three changes of 25 mL celloidin solution in stoppered tubes.

Soak in three changes of toluene: 8 h each.

Impregnate with paraffin wax: two changes, 2 h each.

Embed and cut as usual.

6.3   CUTTING SECTIONS

Cutting paraffin sections is often regarded by many as a dark art, and as such best left to skilled technical staff. It is certainly true that expertise is more commonly found amongst technicians than amongst research workers. The problem for the research worker who wants to cut his or her own sections is that skilled technicians will nearly always make a better job of it than he or she can, at least at first. This can lead to discouragement. Of course, if the research worker is working without expert technical assistance, there is little alternative to learning to cut their own material. In any case, knowing how to cut sections is an important part of the training of any research worker who intends to use histological techniques. Losing control of one's own material is undesirable and there are real disadvantages in not being able to follow one's own material through the entire process from post-mortem examination to stained sections. Shrinkage needs to be assessed if quantitative techniques are to be used, which means that blocks have to be examined, indeed measured, after infiltration with wax and knowing "how the tissue cut" is important in assessing the adequacy of fixation. If the reader is not interested in preparing his or her own sections, he or she could skip this section, but may miss something by doing so.

Three aspects of the process are especially important to understand: the wax, the knife and the microtome. Knowing something about each of these might be thought to be obviously necessary; rather oddly, some workers seem to manage without this knowledge. They use whatever wax the laboratory has always used, they use disposable blades without wondering whether some other blade would be better and, less surprisingly, they use whatever microtome is to hand. Of course, there is nothing wrong in this for the beginner: everybody has to start somewhere. But, as expertise develops, it is at least worth thinking about these things and, perhaps, modifying the methods used. It is certainly a mistake to accept unthinkingly that one's research is constrained by one's methods.

The thinnest sections that can easily be achieved are about 4 μm in thickness and sections thicker than 20 μm can be difficult to manage. For the majority of applications you will want to cut your sections as thin as possible. Such slices cannot be cut using the drawing or sawing action of an ordinary knife, as if one were cutting slices off a loaf of bread, but can be cut by paring the sections from the block. This requires careful control of the cutting process, which is where the microtome comes in. It is a precision instrument made to work to very fine tolerances and needs to be properly maintained and used carefully. In essence there are two types of microtome: those in which, during the cutting stroke, the block is held still and the knife moved across it so as to remove very thin sections, and those in which the knife is held still and the block moved across its edge. But in both cases the knife acts like a carpenter's plane and at each stroke it planes a very thin shaving, a section, from the block. Between each cutting stroke the block is moved forward so as to allow the next section to be cut at precisely the same thickness as the last. For those who like homely analogies: imagine a banana being slowly pushed up through a hole in a table and a knife, skimming back and forth over the surface of the table, cutting off thin slices of the fruit. This, by the way, is exactly how early hand-microtomes worked. Microtomes are made so as to allow the advance of the block to be controlled to within a micron; this is why they are expensive.

6.3.1 Basic Types of Microtome

6.3.1.1 Microtomes in which the Block Moves and the Knife is Fixed

Rotary microtomes

The rotary microtome is now in general use in histology laboratories (Figure 6.3). The block, fixed to the chuck, moves up and down across the edge of the fixed blade. Between each cutting stroke, the block moves forward by the pre-set distance: the thickness of the section. Each section adheres to the last and a ribbon is formed.

Figure 6.3 **Standard laboratory rotary microtome.** No ''bells and whistles'' on this one, you just turn the handle and the chuck moves up and down.

Because the block moves perfectly vertically, the sections are of perfectly even thickness from edge to edge, are parallel with each other and are very suitable for reconstruction work: just like the slices of a sliced loaf of bread. The modern rotary microtome is often power-driven, the speed being controlled by a foot pedal as in the case of an electric sewing machine. The force applied to each cutting stroke is perfectly controlled and such microtomes, though expensive, are very efficient. Power-driven microtomes can be used for cutting plastic sections; all microtomes used for cutting resin sections for electron microscopy (EM) are power-driven. Despite these advantages there is something to be said for using hand power: one can feel the cutting stroke and adjust the power applied to down-stroke accordingly, and many experienced technicians prefer the extra control and feel they get from turning the handle themselves.

The Cambridge rocking microtome

The reader might come across the Cambridge rocking microtome. This splendid old instrument worked by causing an arm to rock on a horizontal support. The end of the arm carrying the chuck and the block moved up and down in an arc. At each down-stroke, powered by a strong spring, the block passed across the edge of the blade and a section was cut. The arm was then raised by moving the handle and the next down-stroke followed. The rocking arm was automatically advanced between each cutting stroke. This microtome is as good as a rotary microtome for preparing ribbons of sections but, and this is the reason for preferring a rotary microtome, the sections are cut on an arc at 90° to the blade and not vertically. In practice this makes very little difference unless reconstruction work, for example, as needed in embryology, is being undertaken. The rotary microtome was, in fact, invented by Minot for such work. For ordinary paraffin wax sectioning there is nothing wrong with the Cambridge rocking microtome, though it is a little more difficult to learn to use than the rotary. Learning to click the handle to and fro at an even and appropriate pace seems more difficult than learning to turn the handle of a rotary microtome. The "rocker" is cheaper than the rotary. There used to be a second type of Cambridge rocker in which the block moved up and down parallel with, and not at 90° to, the blade. This version did cut perfectly flat and parallel sections and could be used for demanding embryological reconstruction studies. To the best of our knowledge, this machine was not as widely used as the standard Cambridge rocker described above and we have never seen one in use. The standard model was called the "small rocker"; the alternative model was the "large rocker".

The sledge microtome

With the sledge microtome the block is moved to and fro on runners or rails, and the knife is fixed so that, at each stroke, a section is cut from the surface of the block. The block rises by the pre-set distance before each forward-stroke. The knife can be angled across the line of travel of the block. Sledge microtomes are big, heavy instruments and, because of this, are very effective for cutting tough tissue. Their one drawback is that it is not easy to produce a ribbon of sections, although this can be done with practice if the knife is set at right angles to the direction of travel of the block. Nobody would choose a sledge microtome for serial sectioning, but many histologists use a sledge microtome for routine preparation of sections.

6.3.1.2   Microtomes in which the Knife Moves and the Block is Fixed

Sliding microtome

The sliding microtome differs from the sledge in that the block, carried on its chuck, rises at each stroke and it is the knife that is moved horizontally. The knife can be angled. This instrument is mainly used for cutting celloidin sections. For cutting celloidin sections, it is exceptionally good. An attachment for producing freezing sections can be added.

The bench-mounted freezing microtome (sometimes called the "clinical microtome")

Here too, the knife is moved horizontally across the block. The block rises before each stroke of the knife. Note that, with all the instruments described above, the block is advanced between each stroke. The knife is either held still (rotary, rocker, sledge) or moved in a horizontal plane (sliding, freezing).

6.3.2   Microtome Knives

It goes without saying that the knives used for cutting sections need to be sharp. In fact, they need to be as sharp as possible and sharpening one's own microtome knife used to be a test of one's ability as a microtomist. The reader may be glad to know that the days of Belgian Rock hones, of glass plates and aluminium oxide grinding powders, of horsehide strops and sticks of grinding compound have passed. In fact, even if non-disposable microtome knives are used, they are likely to be sharpened by automatic sharpening machines rather than by hand. It is now generally accepted that an automatic sharpening machine can produce just as good an edge in much less time and with minimal effort, with the result that hand-sharpening is now largely a thing of the past. The process of sharpening can be divided into two parts: setting the bevel and polishing the facets of the bevel. Automatic sharpeners do both: a finer grain abrasive paste is used for polishing than for setting the bevel. For anyone interested in the lost art of hand-sharpening, excellent accounts are provided in the older manual of technique: see Bibliography.

Many microtomists now use disposable blades instead of the older large and heavy microtome knives. Manufacturers provide special holders that allow the disposable blades to be clamped at the correct angle for cutting sections. Disposable blades work perfectly well for ordinary section cutting; only if very large sections are being cut or, perhaps, if the material is unusually tough might one turn to an older fashioned microtome knife. It should be remembered that there are few things in a histology laboratory more dangerous than a heavy microtome knife with an edge a foot long and as sharp as a cut-throat razor. Handling such a knife requires care: serious injuries can result from careless handling. This is particularly important when screwing the handle into the end of a long knife. Heiffor knives, as were generally used with the Cambridge rocking microtome, had the handle permanently attached to the blade and were a little safer. One piece of good advice: if you drop a knife NEVER try to catch it with your hand or break its fall with your foot! If old-fashioned knives are in use then each user should have and be responsible for their own knife.

6.3.3 Tilt and Clearance

The angle of tilt and the clearance angle are shown in Figure 6.4. If the clearance angle is not sufficient to allow the back surface of the knife to pass clear of the front surface of the block, it will drag up the block producing compression of the wax and tissue. The next section will then either not be cut or will be cut too thinly, and will probably break or be carried up on the block as it moves away from the edge of the blade. Wrinkling or rolling up of sections is a sign that the clearance angle is too large. The tilt of the knife controls the clearance angle: for soft tissues the tilt can be larger than for tougher tissues. There is no way of knowing in advance the perfect tilt angle for any particular block: only practice and experience can teach this. It is sensible to begin with the knife set at about 20° to the vertical and then to increase or decrease the tilt after a few sections have been cut. Of course, if a perfect ribbon is produced at once, then leave the knife where it is! The thin disposable knives allow the knife to be used in a more upright position than the older knives, which had thick backs: as shown in Figure 6.4.

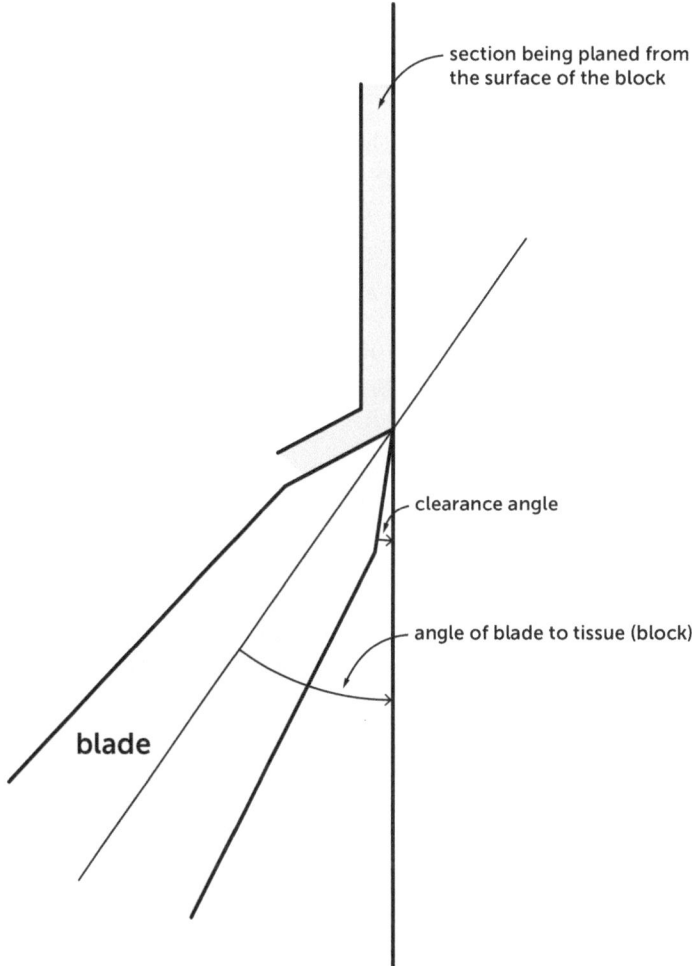

Figure 6.4 **A block being cut by a microtome knife.**

6.3.3.1 Angle. The knife can also be angled with some microtomes. This means that the edge of the knife meets the block by installments and not all at once. Angling is standard with sledge and sliding microtomes but is not used with a rotary microtome. Angling is probably impossible with a rocking microtome.

6.3.4 Starting to Cut: Using a Rotary Microtome

There is much to be said for learning how to cut by cutting a wax block that does not contain any tissue. Not many people bother to do this and the following description assumes you are cutting "for real".

Begin **without** the knife in place. If you are using a system that produces a block in a holder, place the block between the horizontal jaws of the chuck and tighten it into position. Retract the mechanism so that, on turning the handle, the block passes up and down at some distance from where the edge of the knife is going to be, half a centimeter should be enough. The last thing you want is for the block to be sticking out over the edge of the knife and to smash down on the edge when you make the first cutting stroke. Getting to know the microtome is essential before you start cutting: practice putting the knife in place and look at the relationship between the knife and the chuck.

To produce a perfect ribbon of sections, the top and bottom edges of the block must be parallel to one another. How the tissue lies in the block doesn't matter at all as far as forming a ribbon is concerned. This is easy if a disposable system for embedding is used, the top and bottom edges of the carrier of the block will be parallel to the edges of the block, and will be parallel to the knife. If the block has been attached to a piece of wood or metal and that has been placed in the chuck then things are a little more complicated. You will have needed to pare the upper and lower surfaces of the block until they are parallel and then set the block holder so that these top and bottom surfaces are parallel with the edge of the knife. This all sounds horribly complicated; in fact, it is not difficult and a lesson from an accomplished technician will show you how to do this in a few minutes.

Assuming you are right-handed, the microtome should be put on the bench with the handle towards you. You will be turning the handle with your right hand. I have not seen a left-handed microtome, so if you are left-handed you should learn to use your right hand to turn the handle. However, if you are left-handed, you will have an advantage: your left hand has the more difficult task. Back to the right-handed person: your left hand will be used to manipulate the ribbon. As the sections are cut, they will form a ribbon, which will fall down the side of the blade away from the block. If you didn't pick up the end of the ribbon, it would simply pour down onto the bench as you went on cutting.

Put the knife in place and lock it in position using the locking screws. The knife must be firmly locked in place: no movement of the blade can be allowed during cutting. Now set the section thickness control to 8 μm and turn the handle. Of course, the block is too far from the blade for any sections to be cut. Keep turning the handle: the block will advance towards the blade: watch the gap closely. If a fast advance control is available you can use that to move the block towards the knife, but make sure that you stop short of the knife. As the block begins to strike the blade, sections will appear. At first there will be no tissue in the sections as the layer of wax that was beneath the tissue when you embedded it is now being cut. These wax-only sections should form a nice ribbon. As soon as the tissue appears in the section, stop with the chuck at the top of its travel. Keeping your right hand on the handle, brush away the wax-only sections from the knife with your left hand. It is important that no wax should remain on the knife edge. If some wax does remain, apparently stuck on the edge, do not try to push it off with your fingers if you value your fingertips; use a fine camel-hair brush instead.

Once the surplus wax is removed, you can begin to turn the handle and cut your sections. As the ribbon begins to form, pick up its trailing end with a camel-hair brush. Dampening the brush helps the ribbon to stick to its hairs. The brush is, of course, in your left hand. Raise the end of the ribbon a little so that it hangs in a curve from the knife. Go on cutting. The ribbon elongates and you will need to move your left hand and the brush away from the blade to keep the ribbon hanging in a nice curve. There is something curiously satisfying about cutting a ribbon: turning the handle, manipulating the brush and seeing your sections forming (Figure 6.5). As soon as you have a ribbon of about a dozen sections, stop cutting with the chuck at the top of its travel. Now, with your right hand, pick up another brush and

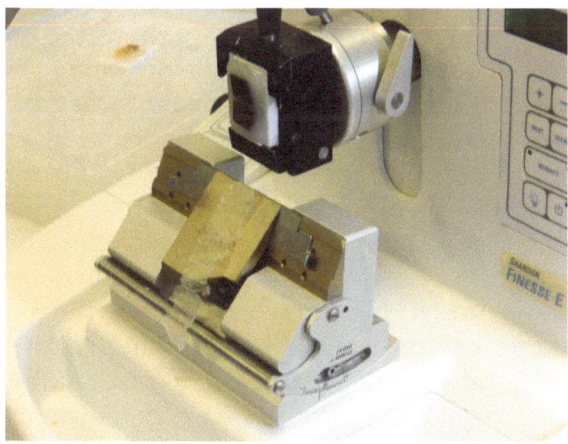

Figure 6.5 Microtome with a block in the chuck and a ribbon coming off.

remove the last section from the blade with a delicate upward motion of the brush. It will part company with the blade easily. The ribbon is now supported by the two brushes. Move your hands and put the ribbon down on a sheet of paper. If you have some black paper that will be better than white: it's easier to see the ribbon on the black surface. Shiny surfaced paper used to be recommended, but plain white paper is perfectly satisfactory. You have a ribbon! Don't lose it by allowing it to blow about in the wind! Ribbons are best placed in a shallow box about 2 cm deep that can be covered with a sheet of glass. I prefer glass to a sheet of Perspex: the latter can acquire a charge of static electricity and the ribbon can be attracted to it.

Cut another ribbon and lay it alongside the first. Now stop. You have cut more than enough sections to begin with. Wind back the mechanism that advances the block, remove the knife, remove the block and adjourn for coffee. The next stage involves attaching the sections to slides. We shall come to that shortly.

6.3.5 Dodges and Tricks of the Trade

Standing the block on ice before cutting is helpful: the face that you intend to cut should be in contact with the ice. Immersing the block in cold running water is also a helpful dodge. There seems to be more to the effect of water than simply cooling. Do not leave the block in water for too long: the tissue will take up water, expand and stick out from the surface of the block.

Breathing on the sections exposes them to water vapour and this reduces the tendency of the wax to pick up a charge of static electricity and to lift from the knife.

If the sections tend to curl up as they are cut, use the brush to hold them flat. You might be wondering whether this calls for a third brush and a third hand. Use your left hand. Once the ribbon has begun to form you will not need to worry about sections curling.

Give a little more pressure on the down-stroke of the blade than on the up-stroke: a positive down-stroke is important.

You may find that picking up the ribbon with the dampened tip of your left index finger is easier than using a brush. Some experts recommend using forceps, but this can damage the section that is gripped. However you do it, handle the ribbon gently. In the great days of histological technique, microtomes were fitted with little conveyer belts that collected the ribbon. Yet more impressive were drums upon which the ribbons were wound: the drum turned with each stroke of the microtome. Such refinements are not seen today.

Always put the ribbon down with its shiny side to the paper. The shiny side is the side that will be placed on the glass when the sections are fixed on the slides.

Beveling one of the top edges of the block is useful in that it allows you to tell one side of ribbon from the other. This is helpful when mounting serial sections for reconstruction work.

6.3.6 When Things go Wrong

Sooner or later, you will have problems: everybody does. The manuals of technique provide long lists of problems and propose solutions. The list given in Table 6.1 reflects the author's experience and reading: consult the books listed in the Bibliography for further advice.

Most of these problems can be solved by:

- Ensuring that the tissue is properly fixed, dehydrated, cleared and embedded;
- Ensuring that the knife is sharp;
- Ensuring that the tilt of the knife is appropriate.

6.3.6.1 What to do if the Sections Crumble and all Adjustments of the Knife, Changing the Knife and Cooling the Block Fail to Produce Decent Sections. Much now depends on how important the particular block is to your research. If you have other blocks, set aside the difficult block and cut one of those. If you have tissue lying in fixative, process some of this very carefully and try sectioning again. Poor processing is a common enough cause of sections crumbling and something might have gone awry. Consider double embedding the material: this is strongly recommended for tissue that is difficult to cut because the celloidin offers much firmer support than paraffin wax.

If the difficult block is critically important and can neither be discarded nor replaced then proceed as follows. Take the block and place it in absolutely clean molten paraffin wax in the wax oven. It may be that the clearing agent had not been completely removed in the earlier infiltration sequence. Better: put the block in a bath of molten paraffin wax in the vacuum embedding apparatus for a couple of hours: this will improve the re-infiltration and speed up the process (see section 5.10.3.2). Cast a new block after re-infiltration. Now, using a sledge microtome (not the rotary), attach the block to the carrier and try cutting single sections. Begin with the knife set to about 20° but vary that if you don't get good sections straight away. If re-infiltration has solved the problem, single sections will be cut without difficulty.

Table 6.1 Section cutting: problems and their solutions.

Problem	Cause	Remedy
Ribbon is curved	Top and bottom of the block are not parallel	Trim block (unlikely to be necessary with modern blocking methods)
	Knife not uniformly sharp	Change blade or sharpen knife
Sections compressed: not as wide as the block	Blunt knife	Change blade or sharpen knife
	Wax too soft or room too warm	Stand block on ice for 30 minutes. Some compression is inevitable and will be sorted out during fixing the sections to the slides
Sections alternate between thicker and thinner	Knife loose	Tighten knife in holder: ensuring that all holding screws are tight before you begin to cut is important. If the knife is not firmly held then the sections may show the artefact of chattering.
	Too much tilt of the knife	Reduce the tilt of the blade. Tough tissue cut with too much tilt screams as it is cut
Sections split or scratched longitudinally	Nick in the knife	Change blade or sharpen knife
	Grit in the object	The grit may be bone that has not been adequately decalcified. If you are cutting serial sections and the particular block is critically important, this is a real nuisance. All you can do is look for the grit and pick it out from the face of the block with a needle. You will have ruined some sections, of course. Better to start again and improve decalcification
Sections roll up and will not form a ribbon	Too much tilt	Reduce tilt of knife. With too large an angle you are scraping not cutting the wax.
	Wax too hard	Stop cutting. Find some low-melting point wax and melt some in a beaker. Paint the top surface of the block with the molten wax and allow it to set. Start cutting again. Warm the room: put on a heater. The old advice to "stir the fire" (coal fires in laboratories!) is no longer relevant

Observation	Cause	Remedy
Ribbon lifts with the block instead of lying on the blade	Effect of static electricity	This can be a real nuisance. Ribbons can fly about and stick to block, to the metal of the microtome and curl into fantastic shapes. How to reduce static? Several methods have been suggested: – Breathe on the sections – Boil some water near the microtome: the water vapour picks up the charge – Light a Bunsen burner near the microtome – Don't wear rubber-soled shoes – Employ an ioniser in the lab – Wait for a more humid day
	Bits of wax stuck on the edge of the knife	These bits have to be removed. Wipe blade upwards (NEVER ALONG THE EDGE) with fingertip. Be very careful of both your finger and of damaging the edge! Remove with finger tip
	Bits of wax stuck on the top edge of the block	
Sections look OK but won't ribbon	Wax too hard	Warm the room or apply softer wax to the surface of block: see above. Cutting a little faster often helps
Feeling of resistance on the down-stroke of the handle, next section may not cut or be cut too thin	Not enough clearance: the block is hitting the side of the knife and being compressed	Increase the tilt of the knife

Angle the knife a little so that it does not strike the block "square on". If the sections still crumble, move to the next stage.

Make up a solution of celloidin: 1% in a 50/50 mixture of absolute alcohol and ether. Remember to be careful when handling ether: no naked flames! Paint the surface of the block with the celloidin solution using a brush. Allow just a few seconds for the celloidin to dry and cut a section. The celloidin should hold the tissue together. Pick up this valuable section with a brush moistened with albumin adhesive (not the one you used for the celloidin) and transfer it to a perfectly clean slide that has been flooded with albumin adhesive. Place the slide on a hot plate (45–50 °C) and allow the section to flatten. Flattening is not as easy as with ordinary paraffin wax sections: the celloidin prevents it. If the section will not flatten you will have to warm it a little more: this can be done over a small flame (remember the ether: close the bottle and put it away after use). It is important *not* to heat the section to *above* the melting point of the wax, the aim is to heat to just *below* the melting point of the wax so that it will expand and flatten nicely. As soon as the section has flattened, wipe excess fluid from the edges of the section and put it in the 37 °C incubator to dry. Dry it well, this is a valuable slide! Repeat until you have enough sections for staining. All rather tricky and tedious and only necessary if you really cannot afford to discard the block! Always consider reprocessing but never actually throw your bad blocks away no matter how tempting this may seem: keep them as you may need to attack them again!

6.4   ALTERNATIVE EMBEDDING SOLUTIONS: CELLOIDIN AND FROZEN SECTIONS

6.4.1   Celloidin Sections

Celloidin is the name generally given by histologists to nitrocellulose. Celloidin is distinguished from low-viscosity nitrocellulose (LVN), but they are both forms of nitrocellulose. Histologists speak of "celloidin sections" whether celloidin itself or LVN has been used in their preparation. LVN has the advantages that it can be prepared in higher concentrations, it penetrates more rapidly into tissue because of its lower viscosity and the blocks that are produced are harder than in the case of celloidin. The LVN technique is recommended.

Celloidin, meaning celloidin or LVN, has both advantages and disadvantages when compared with paraffin wax as an embedding medium. Briefly:

Celloidin is a much tougher embedding medium than paraffin wax. Larger blocks can be used and the sections can be handled easily: they are not fragile. This allows the sections to be handled and stained before being attached to a slide. The celloidin embedding process is carried out at room temperature: the effects of placing the tissue in molten paraffin wax are avoided and shrinkage and distortion of the tissue are minimised. This is very valuable when sections of brain are being prepared. It is also helpful, as pointed out in all the manuals of technique (see Bibliography), when large sections of such a demanding object as the eye are being prepared: keeping the retina, choroid and sclera close together presents a significant technical challenge if wax is used because of shrinkage, but it can be achieved by embedding in celloidin. Celloidin, being itself tough, is a better medium than wax for embedding tough tissues: matching the toughness of the medium to that of the tissue is helpful when sections are cut.

The main disadvantage of celloidin embedding is that it is very slow, taking weeks rather than hours or days for adequate embedding. In addition, celloidin cannot be cut into such thin sections as wax; 15–20 μm is the standard thickness of a celloidin section. This makes celloidin sections less suitable than wax sections for cytology. Finally, celloidin sections cannot be cut as ribbons, they are cut and handled individually, and thus serial sectioning is much more tedious than in the case of wax sections. If, however, you need to use this technique, an outline methodology is given in the following sections.

6.4.1.1 Embedding in LVN. Tissue should be fixed in buffered formalin and dehydrated. Larger blocks may be taken than when wax is being used. Sectioning of a whole rat brain is entirely feasible, indeed very much larger sections can be cut.

Make up the following solutions:

- A 50/50 mixture of absolute alcohol and ether;
- 5% LVN in the alcohol–ether mixture, the solution should contain 1% tricresyl phosphate as a plasticiser.
- 10% LVN in the alcohol–ether mixture, the solution should contain 1% tricresyl phosphate as a plasticiser.
- 20% LVN in the alcohol–ether mixture, the solution should contain 1% tricresyl phosphate as a plasticiser.

Method

All the following treatments should be carried out in closed jars. Transfer the tissue from jar to jar using a pair of smooth broad-tipped forceps or pour out the solution and add the next when a change is made. If you use forceps, don't squeeze the tissue.

Place the tissue in:

The alcohol–ether mixture for	24 h
5% LVN for	5 days
10% LVN for	5 days
20% LVN for	3 days

Leaving the tissue in any of the solutions of LVN for a little longer than suggested above will do no harm.

Whilst the tissue passes through LVN, you have time to make some cardboard boxes of appropriate size. This is a task for a wet afternoon when reading reprints or making statistical calculations fails to appeal. Follow Figure 6.6 and the following description closely:

How to make a box (adapted from Bolles Lee)

Always use thin cardboard: plain postcards are ideal. Always score the lines with the point of a pair of scissors before making the folds.

Having marked out the lines and scored them: Fold along the lines a–a′ and b–b′ and then along c–c′ and d–d′. Always fold the same way. Fold along A-A′, B-B′,C-C′ and D-D′. This is remarkably easy if you have scored the card. Those with good spatial skills will now see the box forming. Fold along A′–B′ and C′–D′ and turn the dog ears round the corners. Note that folds A′–B′ and C′–D′ go the other way from the original folds. Use sticky tape to hold the dog ears in position. Having made one box, you can now make five more. The depth of the box is, of course, determined by the distances between c and A and c′ and A′: these must be identical to one another. Make a few boxes of different sizes. Children can be drafted to help with this.

Ensure that, when the tissue is placed at the centre of the box, there will be 1 cm or so of celloidin all around it.

Pour a little 20% LVN into the cardboard box. Place the tissue in the box, the face from which you intend to begin cutting sections should be towards the bottom of the box. Fill up the box with 20% LVN.

Place a small beaker of ether in the bottom of a large desiccator, replace the perforated shelf and put the box of LVN on the shelf. Replace the lid. The ether aids the dispersion of bubbles that may have formed whilst the LVN was being poured.

Once any bubbles have disappeared, after about an hour, replace the beaker of ether with one containing chloroform. The block will harden. This takes about three days. Top up the chloroform as it evaporates.

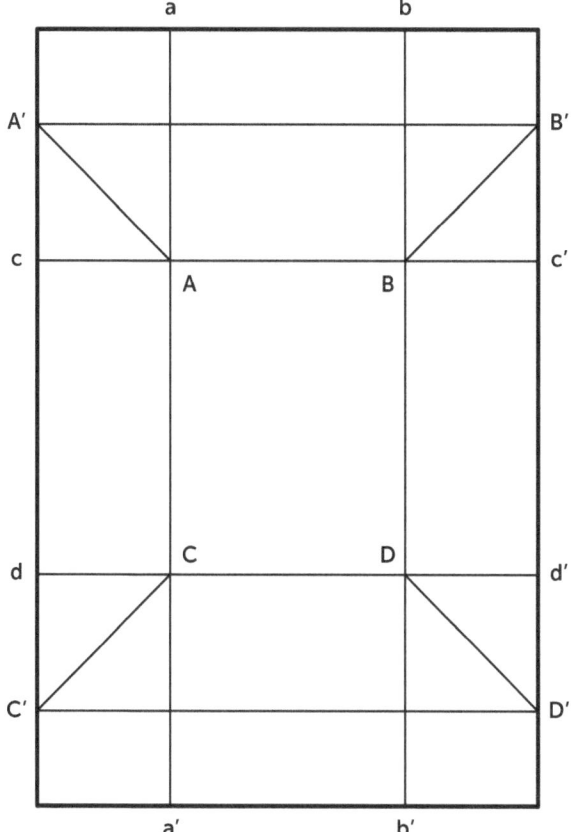

Figure 6.6   **Schematic of how to make a cardboard box for embedding in LVN** (refer to the method described in the text).

Remove the box and peel off the cardboard.
The block will have the consistency of hard rubber.
Store the block in 70% alcohol.

6.4.1.2 Cutting Celloidin Sections.   Celloidin sections are easier to cut than sections of paraffin wax. Celloidin sections are cut on a sledge or sliding microtome. The firmness of the block and its, often, large size seem to call for a standard microtome knife and it seems probable that the disposable blades that are the staple of the rotary microtome would likely be insufficiently rigid. The knife must, of course, be sharp. Sharpening using a sharpening machine is recommended. Remember to use a polishing disc or polishing abrasive for the final period on the machine. Sledge microtomes are in wider use than sliding microtomes and the following description assumes that you will be using a sledge. Remember that, with the sledge, the knife stays still and the block is moved to and fro and rises at each stroke. In the case of the sliding microtome, the knife is moved to and fro and the block rises at each stroke. On sliding microtomes the block carrier climbs an angled screw thread: this means that there is some horizontal travel in addition to the vertical movement.

The block must be firmly attached to the block-carrier. The block-carrier should be larger than the block and made of wood or plastic. The block is attached by seating it in a small pool of 20% LVN and allowing it to harden. This takes an hour or so. Putting the carrier plus the block back in the desiccator of chloroform vapour will aid hardening.

Once hardened, place the carrier in the chuck of the microtome and adjust the height of the chuck so that the upper surface of the block is well below the plane of travel of the knife edge.

Place the knife in the knife holder and set the slant angle to about 25°. Adjust the tilt angle to about 20° and tighten all the screws that hold the blade in place.

Move the block towards the knife. Move it slowly and watch carefully: the upper surface of the block should pass beneath the edge of the knife.

Set the section thickness: 20 µm is appropriate.

Using a pipette, wet the knife and the upper surface of the block with 70% alcohol. An automatic drip-feed of 70% alcohol can be arranged with a raised bottle, some plastic tubing and an adjustable clamp: this is not needed if you intend to cut only a limited number of sections.

Slide the block to and fro noting the click as the mechanism raises the block after each back-stroke. Make sure that you draw the slide that moves the block fully back on each stroke: this is necessary to activate the lifting mechanism.

As sections begin to appear, they will be deposited on the upper surface of the knife. Remove them using a brush dipped in 70% alcohol or a pair of forceps. Though thin, the sections are quite robust and can be safely handled with forceps. Place the sections in flat dishes containing 70% alcohol. If you want to put more than one section in a dish, and you want to keep them in order (as if you were preparing serial sections), place a circular piece of filter paper on top of each section after you put it in the dish.

Re-wet the block and knife between the cutting of each section.

Once you have enough sections, remove the knife, dry it carefully and put it away in its proper box. Now remove the block and carrier. If you do not plan to cut any more sections from the block, separate the block from the carrier using a scalpel. Store the block in 70% alcohol in a closed jar. Dry LVN should be peeled off the block-carrier. Any residual LVN can be removed from the block-carrier by immersing it in the 50/50 alcohol and ether mixture. If the block-carrier is scored or is wood with a saw-cut surface (a help when attaching a block) then cleaning out the score lines is a sensible step. If you do plan to cut more sections from the block then store the block and the carrier together in 70% alcohol.

The sections may be stained as loose sections or attached to slides and then stained.

6.4.2   Frozen Sections

Frozen sections may be cut from fresh tissue or from fixed tissue. In the first case the use of fixatives and embedding media, such as paraffin wax, are avoided.

Drury and Wallington listed a number of advantages of frozen sectioning over conventional (for example, paraffin wax) sectioning:

- Speed may be of the essence in some situations. For this reason, frozen sections is the method of choice for examining many biopsies in National Health Service (NHS) laboratories. If the tissue looks malignant, a surgeon can make a rapid decision with regards to more radical excision.
- Preservation of lipids, which are, in general, removed by conventional processing.
- Preservation of enzyme activity and certain epitopes (some cluster designation (CD) markers are very heat labile), which would be reduced both by fixation and by processing in hot paraffin wax.
- Avoidance of the artefacts that are always seen after paraffin wax processing of tissue from the central nervous system.
- Immuno-fluorescent antibody techniques are possible, although there are now plenty of antibodies available that work on formalin-fixed paraffin sections.
- Preparation of sections of tissue that are difficult to cut in paraffin wax, for example, tendon. Paraffin sections of such tissues are, in addition, difficult to flatten on slides.

There are a range of histological techniques that involve the use of low temperatures and these are often lumped together under the "frozen sections" banner, which can be confusing. The following annotated classification may be helpful.

A. Fresh frozen sections

No pre-sectioning fixation is involved; the tissue is frozen to the tissue carrier and cut. Sections can be cut either on a freezing microtome or using a cryostat.

   a. Sections can be mounted on slides and stained directly from the microtome;
   b. Sections can be post-fixed in chilled (4 °C) calcium formalin or acetone and then mounted on slides for staining.

B. Frozen sections of fixed tissue

Sections can be cut from tissue that has been boiled for a few minutes either in water or in 10% formalin. These are generally used in diagnostic work. Fixed tissue is frozen onto the tissue carrier and cut either with a freezing microtome or using a cryostat.

Tissue can be fixed in 30% sucrose in 10% formalin and then frozen and cut. This is especially helpful in cutting sections of brain.

The fixed tissue can be impregnated with gelatin, allowed to solidify and then cut, again either on a freezing microtome or using a cryostat.

Routinely fixed tissues can be cut frozen so as to preserve the presence of soluble entities, such as lipids or glucose

C. Freeze-drying

The idea of freeze-drying is to remove all the water from the tissue by rapidly cooling it to very low temperatures, for example, -150 °C in isopentane cooled in liquid nitrogen, then allowing it to warm. As the tissue warms, the water is sublimated. Thus, entirely dry tissue is produced. This can then be embedded in paraffin wax or resin. This process avoids the use of fixatives and also of dehydrating agents, such as alcohol. Sections are cut in the usual way. Note that the sections are not fixed: special techniques for handling them and for storing the blocks are required: see Drury and Wallington. Freeze-drying has been largely replaced by the use of fresh frozen sections.

D. Freeze substitution

Here, the tissue is cooled as in the freeze-drying process and the water (actually ice, of course) is replaced with absolute alcohol or acetone, which is cooled to -60 to -70 °C. The tissue is then returned to room temperature and embedded in paraffin wax or resin. This process also avoids the dehydrating agents. Sections may be cut and handled in the usual way and blocks simply stored like standard paraffin blocks.

Deciding on which technique to use depends on what you are interested in studying. If you are interested in enzymes then fresh frozen sections will appeal, but do check if you can't perform your experiment on formalin-fixed sections first as it is significantly easier than frozen sections and the tissue architecture is better preserved.

If you are interested in the histology of the central nervous system (CNS) then frozen sections of fixed material may be appealing. Drury and Wallington provide excellent advice in their long chapter on techniques applicable to the CNS. In selecting a freezing technique you should decide what you really want to *avoid*: the effects of fixatives? The effects of dehydrating agents? The effects of exposure to hot paraffin wax?

6.4.2.1 *Microtomy with Frozen Sections.*

Sections of frozen material may be cut using either a freezing microtome or a cryostat. The freezing microtome is mounted on the bench and the tissue frozen with liquid carbon dioxide, which is delivered from a cylinder to the tissue carrier. The knife blade should also be cooled. Thermo-electric cooling of the block and knife is also possible. Solid carbon dioxide may be used to freeze the block to the carrier before this is placed in the cryostat chamber. By contrast, the cryostat encloses the entire microtome at low temperature (-5 to -30 °C).

6.4.2.2 *Cutting Sections Using a Standard Bench-Mounted Freezing Microtome or a Sliding Microtome.*

The bench-mounted freezing microtome has, to a large extent, been replaced by the cryostat. But cryostats are expensive and the reader might not have access to one. Good work can be done using the bench-mounted freezing microtome, such as the Reichert-Jung sliding microtome with a freezing stage to cut frozen sections. This excellent instrument can be used for cutting celloidin sections, or indeed paraffin wax sections, as well as frozen sections. The Leitz freezing microtome, which used to be known as a clinical microtome because of its use in hospital laboratories for rapid diagnostic work, is also a very satisfactory instrument, but is not used for anything other than cutting frozen sections. In both, the block stays still, except for rising between strokes, and the blade is moved along rails. Sledge microtomes can be used but, because the block moves, so must the connection between the freezing stage and the carbon dioxide cylinder and a longer connecting tube will be needed: see below.

The block is cooled by liquid carbon dioxide flowing from a cylinder into the sub-stage compartment. When the controlling valve is open, the carbon dioxide vaporises and the vapour escapes through holes around the stage. Liquid carbon dioxide is also directed onto the blade. Carleton (1957) pointed out that cylinders of carbon dioxide are under high pressure and, when the valve of the cylinder is open, as it is during use, the lever that controls the flow of liquid carbon dioxide to the stage takes the pressure. When this valve is closed, the tubing between the cylinder and the stage takes the strain and must be short and strong. A cylinder with an internal tube should be used; it should be mounted beneath the microtome. (Drury and Wallington point out that, if a cylinder without a tube is used, then some carbon dioxide should be blown off before connecting to the microtome. This is to blow off water that may be present in the cylinder and to avoid it freezing and blocking the connecting tube: very good advice!)

Method

Set up the microtome and test the supply of carbon dioxide: ensure that your cylinder is full. Open the valve on the cylinder and let through some bursts of carbon dioxide by opening and closing the valve on the microtome with the lever provided.

Place a small piece of wet filter paper on the tissue carrier and put the block on top of it.

Put the knife in the holder and tighten the screws. With the sliding microtome, the knife can be angled at about 20°; the tilt should also be set at about 20°. With the ordinary freezing microtome (for example, the Leitz model), the knife meets the block square on.

Bring forward the knife and raise the tissue carrier until the top of the block is just below the edge of the blade. Observe from a bending position.

Operate the lever that controls the flow of carbon dioxide. Use short bursts of a few seconds' duration and note that the block turns white and freezes from the bottom upwards. If your instrument has a separate control for cooling the knife, operate that too. You are just about ready to cut.

Sections are cut when the block has warmed slightly from the deep-frozen state induced by the carbon dioxide. At first, the block will be too hard to cut. If you try to cut sections, you will notice that a harsh grinding sound is produced and that the tissue does not produce sections: it comes off as fragments. Judging the temperature at which to cut is the key skill in using the freezing microtome. The tissue may be warmed by putting your finger tip on the top of the block; alternatively, wait a few moments. One time-honoured method, shown to

RLM by Mr Les Jones of Cardiff University, is to tap the top of the block with the blade of a scalpel (the ordinary Swan Morton type). When the tissue is too hard to cut, it gives off a sharp "ping". As the tissue warms, the "ping" changes (its pitch drops to a "pong" you could say) and then the tissue is ready for cutting. As soon as you think the tissue is just right, cut a few sections. Move the blade confidently. You will need to refreeze after every few sections. If you cut when the block is too warm, the sections will come off as pulp and you may knock the block off the stage.

The sections build up on the knife; they do not, of course, form a ribbon. Remove the sections from the knife using a brush that has been dampened with water. Place the sections in a small bowl of water. The sections may be stained as loose sections or mounted on slides before staining. Sections are removed from the water onto slides that have been coated with albumin adhesive, as described in section 6.5 on attaching sections to slides and mounting. Another option is to use positively charged slides, which you can purchase from many major suppliers and do not require any other treatment for helping the sections "stick" to the glass.

6.4.2.3 The Cryostat. Cutting frozen sections with a cryostat is much easier than cutting them with a bench-mounted freezing microtome, as described above. If you have access to a cryostat then you should learn how to use it from an experienced user. Pearse, who invented the form of cryostat in use in many laboratories, provides an outstanding discussion: see Bibliography. The standard laboratory Cryotome™ is essentially a rotary microtome in a freezer cabinet and uses standard replacement microtome blades (Figure 6.7). As indicated above, this is substantially easier to use that the machine described above and, if you need frozen sections, access to one of these will significantly lessen your burden.

6.5 ATTACHING SECTIONS TO SLIDES AND MOUNTING FINISHED SECTIONS

Paraffin wax and frozen sections are attached to slides before staining. Celloidin sections are often stained before being attached to slides or can be attached to slides and then stained. Frozen sections may also be stained "loose" or can be attached to a slide before being stained. Attaching the section to a slide is sometimes referred to as mounting the section. Applying the coverslip to the stained section is also, confusingly, known as mounting, and mounting media, such as

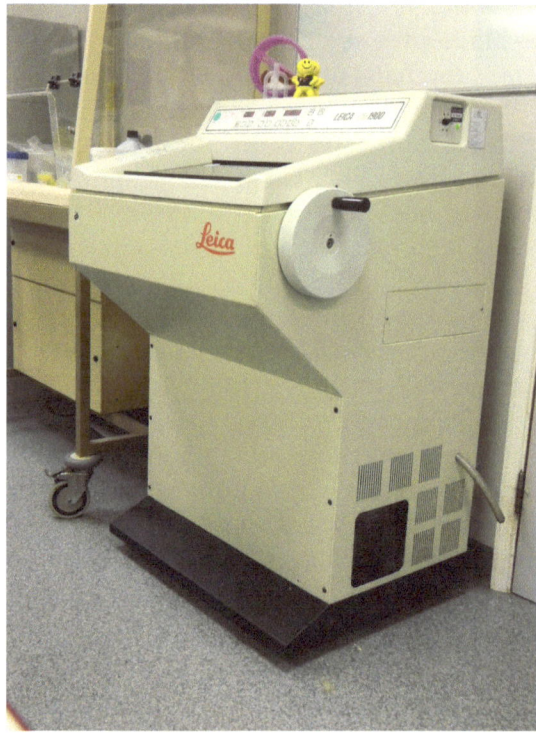

Figure 6.7 **A standard cryostat.** As you can see, this is a far more substantial piece of equipment than the normal microtome. The majority of the extra bulk is due to the refrigeration unit.

distrene–plasticiser–xylene (DPX), are used for this purpose. Here, we shall speak of attaching the sections to the slide and mounting the final stained section.

6.5.1 Attaching Sections to Slides

6.5.1.1 Paraffin Wax Sections. Paraffin wax sections have to be expanded and flattened before or whilst being attached to slides. The sections will have been compressed during the cutting process and will expand to about 120% of their apparent width. Compression occurs in the direction in which the knife passed through the block. This must be remembered when deciding how many sections can be placed on a single slide. Often, only one section is placed on each slide but, if serial sections are being studied, you will wish to place a short series of sections on the slide, or perhaps two rows side by side.

Begin by working with a single section. Sections may also be creased or folded and flattening will be especially necessary.

There are several ways of flattening and expanding sections. Two methods are widely used and are described below. Before you begin, make sure that your slides are completely clean: any grease on the slide will prevent the section attaching firmly to the glass and it may float off during staining. Few things are more annoying than finding important sections floating on the surface of the staining fluid. New slides should be placed in a mixture of 100 mL 95% alcohol and 1 mL concentrated HCl. Then wash in distilled water and polish with a dry rag. Avoid touching the surface of the slide. This is not always necessary and most suppliers provide clean slides.

If you are not using positively charged slides, it is sensible to use an adhesive to attach the sections to the slide. This is especially important if silver stains are to be used: the high pH of the staining solution tends to detach the sections from the slides. Slides can be ordered ready coated with adhesive: if these are available, they should be used. The older manuals of technique describe a splendid variety of adhesives: from albumen solutions to quince jelly and garlic water. Albumen works very well. If you enjoy doing things for yourself, make up a supply of Mayer's albumen adhesive as follows:

For this recipe, mix together 50 mL fresh egg white and 50 mL glycerol, and add 1 g sodium salicylate as a preservative. Allow to stand for a few days and then filter through a loose pad of gauze; alternatively, you can order some from a supplier.

Before starting to attach sections to the slides, prepare a number of slides with adhesive on them. This is easily done by taking a very small amount of adhesive on a clean fingertip and smearing a thin film over the slide. Put the treated slides to one side. You can, if you wish, prepare a lot of slides in this way: store them in a dust-free place, preferably edge-on in a slide box. Some experts advise using a small squeegee (Bolles Lee advised this) rather than the tip of a well washed finger—they argue that epithelial squames can be detached from the finger and turn up under the section, where they can take up stain; although experience has shown that this is a theoretical problem rather than a genuine issue.

Flattening using a water bath

Sections are taken and placed, shiny side down, on the surface of the water in a temperature-controlled water bath set at about 50 °C (Figure 6.8). The temperature should be just below the melting point of the wax. Water baths designed for this purpose are painted black

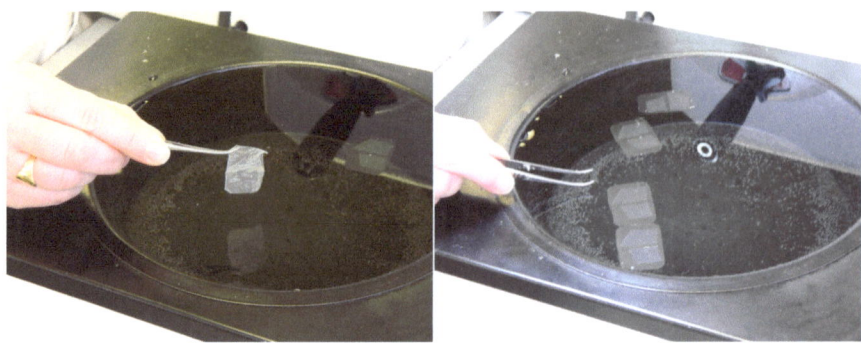

Figure 6.8   **Floating off the sections.** Before (left) and after (right). As you can
imagine, the wrinkled section on the left would be very difficult to do
anything with. The sections flatten out beautifully in seconds on the
water bath. The temperature of your water bath is fairly critical—if you
get it too hot, the sections will disintegrate.

inside to help you see the sections on the surface of the water.
A cautionary tale: the author (RLM) was once given an old water bath.
On using it, he rapidly discovered that the water was "live" and received an unpleasant electric shock. Have your water bath checked
before using it. Checking the temperature of the bath with a ther-
mometer is also a sensible precaution: do this before using the bath
for floating out sections.

Cut a section from the ribbon you have prepared. This should be
done with a scalpel with a curved edge: roll the edge of the
blade across the ribbon. The section may be picked up with a damp
brush. As soon as the section is placed on the water you will see it
expand and flatten out. Take a slide and, holding it vertically, lower it
into the water next to the section. Move the slide so as to pick up
the section. This is easier to do than describe. The lower end of the
slide is angled beneath the section and then drawn upwards carrying
the section with it. A brush may be used to coax the section
onto the slide but, with a little practice, this is usually unnecessary.
Adjust the position of the section on the slide using a fine brush.
Remove the excess water from the edges of the slide with tissue paper
and stand it upright to dry. Drying at more than room temperature is
important: place the slides in a glass-drying oven set at about 37 °C.
Leave overnight to ensure that the section is really dry before staining.

If rumpled sections do not flatten out as well as you had hoped, add
a drop of absolute alcohol to the bath. The section will whiz about on
the surface as a result of the different surface tensions of the water
and the alcohol, but any creases will soon be shaken out. Be careful

though if your section is a bit tattered and torn as it may come apart completely.

Flattening on a hot plate

This method works well for single sections and is better than the water bath for the short ribbons of sections used when working with serial sections. Take a slide and place it on a hot plate set to just below the melting point of the wax. If the wax you are using melts at 56 °C then set the hot plate to 50 °C. Ensure that the hot plate is up to temperature before starting work on the sections.

Examine your ribbon: remember that sections will expand as they flatten so cut out a short series of sections so that the length of the strip is at least one section less than the length of the slide, minus the space for the label. If you cut out a strip of five sections, you will need to leave space on the slide for an extra section or two. You will rapidly discover how much your sections will expand. Using a pipette, place some boiled distilled water on the slide. Using boiled water is helpful as it prevents bubbles appearing under the section. Some people use 70% alcohol instead of distilled water. Lower the strip of sections onto the water and adjust its position. As the water warms, the sections will flatten. Once they are completely flat and you are happy with their position, remove the excess water with a piece of tissue paper and stand the sections in a rack for drying in the oven at about 40 °C. Sections can be also be left to dry on the hot plate. Leaving the sections to dry on the open bench is not recommended, though it often works well enough.

Trouble shooting problems

Sections must be flattened onto the slide and then dried well. Perfect drying is essential if sections are not to come off during staining.

There are two problems that commonly occur. The first is that sections do not flatten out as well as you had hoped. A little judicious stretching of the sections is needed. This can be done by stroking with a fine brush or pulling, very gently, with two pointed cocktail sticks. Be careful: it is easy to break the sections. Davenport advised soaking the sticks in water before use to reduce the likelihood of the points sticking to the wax. The points of the sticks should not touch the tissue, only the wax around the tissue.

The other common problem is bubbles forming under the section. This is unlikely if boiled water has been used—always a sensible precaution. If bubbles do appear, work them to the edges of the sections using a fine brush. Lifting the slide from the hot plate whilst working on the bubbles is helpful; this avoids the section drying down onto the slide with the bubble(s) in place. If the bubble is under the wax away from the tissue, you may prick it with a needle.

6.5.1.2 Frozen Sections.   More care is needed when handling frozen sections as there is no wax present to act as a support for the tissue.

The sections will have been placed in a bowl of distilled or deionized water: they will be floating at the surface. Take a slide that has been thinly coated with albumen and float the section from room-temperature water onto the slide. Take your courage in both hands and blot the section with shiny toilet paper. Put the slides to one side to dry at room temperature.

If you are using a modern Cryotome™, it will be equipped with an anti-roll bar, which will flatten your section out and keep it on the plate within the Cryotome™. Using a slide coated with gelatin, breathe on it gently to warm it a little and then carefully pick up your section off the plate.

Carleton (1957) recommended the following method:

- Place clean slides in 2.5% solution of gelatin in distilled water at 37 °C for a few minutes.
- Wipe one side of the slide clean and allow the other to dry at room temperature, away from dust.
- Float the frozen section onto the gelatinised side of the slide, drain and dry at room temperature.
- When dry, expose to the vapour from 40% formaldehyde for 5 minutes. This can be done in a dessicator: put the formaldehyde at the bottom and put the slides on the shelf.

Carleton pointed out that this method produced excellent adherence and that detachment of the sections from the slides was unlikely even during "complicated staining methods".

6.5.1.3 Celloidin Sections. Celloidin sections are stained free: mounting is needed, which is described in the following section.

6.5.2 Mounting Sections under Coverslips

6.5.2.1 Paraffin Wax Sections.   After staining, paraffin wax sections are mounted in a medium that is miscible with xylene. Many mounting media are available, but DPX is the one that is most widely used and is the one we would recommend. DPX is a solution of distrene (a form of polystyrene) in xylene, with an added plasticiser (originally tricresyl phosphate, now dibutylphthalate). The most famous of the

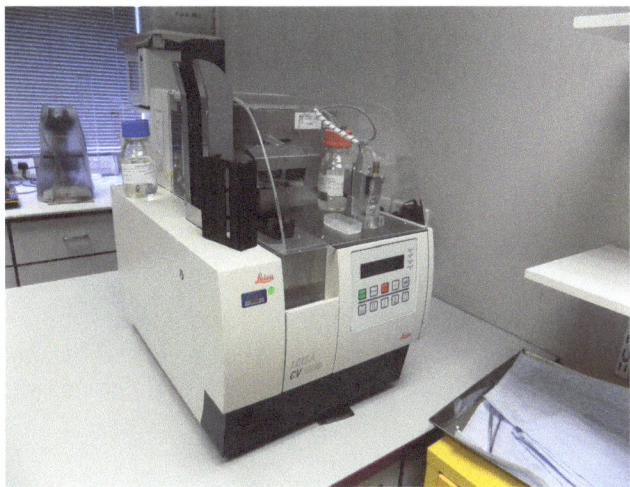

Figure 6.9 Automated cover-slipper.

older media, Canada balsam, can be used, but DPX is better in that it does not darken with age and stains do not fade. DPX has one drawback: it shrinks as it sets and unless a generous amount is allowed to run out from the edges of the coverslip it will retract leaving the edges of the coverslip free.

The principles of mounting are easy to grasp: some mounting medium is placed on the section and a cloverslip lowered on to it. There are several perfectly good methods for doing this, indeed it can be done automatically by a cover-slipping machine (Figure 6.9).

However it is done, the one thing to remember is that the layer of mounting medium between the section and the coverslip must be as thin as possible. If too thick a layer is used then examination with an objective lens with a very short working distance will not be possible: see Chapter 3. It goes without saying that the mounting medium must be free from dust and debris. Moulds can grow in mounting media: the perfectly spherical little black spores of yeasts that multiply by budding are occasionally seen in mounted sections. Xylene evaporates from DPX if the media is kept in an open, or loosely closed, jar. Screw-top jars are better, but the media must not be allowed to accumulate around the thread. Drury and Wallington recommended using a small plastic oil can to dispense the mounting medium: this seems to be excellent advice.

6.5.2.2 Coverslips. These can be obtained in various sizes and thicknesses. The size should obviously be appropriate for the work in hand, and larger coverslips will be needed when serial sections are being mounted. Oblong coverslips are now generally used; the round ones are seldom seen. Thickness is important: 0.13–0.16 mm (Number 1 in the UK terminology) is excellent. Coverslips must be clean and, as coverslips are really not expensive, you shouldn't really consider using anything other than new coverslips. Ensure that they are not exposed to dust. With regard to coverslips, it is worth recalling Carleton's advice:

"If obtained from a reputable firm they should require no cleaning further than that of being kept in hydrochloric acid alcohol until a batch is dried carefully on a grease- and lint-free duster and placed in the Petri dishes [ready for use]. If they are very greasy, treat with potassium dichromate and sulphuric acid mixture [details provided] for a few hours, rinse thoroughly first in running tap-water, then in distilled water. Finally dip in alcohol. Then change your dealer ."

Methods of cover-slipping: number 1
Place the slide on the bench, preferably on a sheet of white paper, and apply a centre-line of mounting medium along its length so that the length of the line equals the length of the coverslip. The line should be significantly longer than the section. Cross the end of the line to form a T shape. The medium is usually applied with a glass rod that has been dipped in the bottle of medium. Don't take too much out on the glass rod. Hold the coverslip in your left hand and allow one end to contact the cross stroke of the T of medium. Lower the free end of the coverslip using an angled dissecting probe held in your right hand. Lower the coverslip slowly so that the medium spreads out nicely beneath it as it approaches the slide. Once the coverslip is down, examine for bubbles. None should be present, but if there are any, press gently on the top of the coverslip with the point of the probe. The medium should have run out from the edges of the coverslip for a distance of 2–5 mm or so. Set aside to dry. This method is needed when large coverslips are being mounted on large slides, such as those used for rows of serial sections. Placing a small weight on the middle of the coverslip to ensure a thin layer of mounting medium is an excellent idea. A 5–10 g weight from an old and unused box of balance weights is heavy enough.

Methods of cover-slipping: number 2
Place the coverslip on the bench, preferably on clean white paper, and apply a large drop of medium. Holding the slide horizontally,

lower it onto the coverslip and let it settle. The weight of the slide presses out the medium. As soon as the medium has flowed from the edges of the coverslip, pick up the slide, invert it and set aside to dry.

Methods of cover-slipping: number 3

Take a coverslip in your left hand and apply a large drop of mounting medium. Invert the coverslip so that the medium hangs from its lower surface and lower onto the slide. This method was recommended by Davenport as a means of minimising the risk of bubbles appearing beneath the coverslip. It relies on the weight of the cover slip to press out the medium: if the medium is rather thick this might not be sufficient and a little pressure may be needed.

A general word of warning with regards to cover-slipping: it may be tempting to try and "improve" the flow qualities of your DPX by adding extra xylene. Avoid this temptation as this dilution invariably leads to increased dry-back and bubbles in the finished slide.

6.5.2.3 Cleaning up the Slides. Do not touch the slides until the DPX has dried completely. It is sensible to wait for a few days because nothing is worse than finding that the coverslip slides and smears the beautifully stained section. Take a sharp pointed scalpel and peel off the excess DPX. It comes off easily. Ensure that you label the slide accurately. Modern slides have ground-glass areas for labelling. Use a fine tipped marker pen with permanent (not water-soluble) ink. If you are a traditionalist, use a sticky label but ensure that it really does stick to the glass. Hyper-traditionalists may like to dip the end of the slide carrying the paper label, with the writing in water-proof ink, in smoking-hot paraffin wax.

6.5.2.4 Frozen Sections. Frozen sections that have been stained on the slide or that have been stained loose and then attached to the slide can be dehydrated and cleared in xylene and mounted in DPX, as described for paraffin sections. BUT this should not be done if the sections have been stained for lipids: these would dissolve in the alcohol. Assuming that you have stained for lipids proceed as follows.

Take the slide from the water and remove the excess water. Do not allow the section to dry or it will curl.

Apply water-soluble mounting medium and put the coverslip in place as described above (Methods of cover-slipping: Number 1).

6.5.2.5 Aqueous Mounting Medium. Aqueous mounting medium is needed if you are staining your tissue sections with alcohol-soluble chromogenic substrates, such as Fast Red. There are many available, some remain in a semi-liquid state and require "ringing" (see below) to prevent leakage from under the coverslip and also to prevent bacteria and mould invading the mountant. The majority of commercially available aqueous mountants set hard and, as such, are easier to deal with.

Lillie's modification of the classic Apathy medium is recommended. This medium has a high refractive index, 1.52, and allows effective use of the oil immersion objective.

Making up the medium

Ingredients:

Pure gum Arabic	50 g
Pure cane sugar	50 g
Distilled water	100 mL
Potassium acetate	50 g
Thymol, as a preservative	0.05 g

Dissolve using gentle heat. Keep in closed jars.

Ringing

It is standard practice to "ring" aqueous mounts. Nail varnish is very effective and, as pointed out by Drury and Wallington, comes in pretty colours and with a brush. Paint round the edge of the coverslip as neatly as you can. In the old days, ringing was an art and turntables were used when ringing circular coverslips. Before experimenting with too many colours of nail varnish, remember Bolles Lee's advice of 1896:

> *"Thanks to the efforts of the dilettanti to outshine one another with neatly gaudy "rings", microscopical literature contains a goodly show of receipts for cements and varnishes. I have collected such as appear likely to be useful, rejecting all that relates merely to ornament."*

6.5.2.6 Mounting Celloidin Sections. The method given in the chapter on staining methods (Chapter 7) is repeated here for convenience: assuming that the sections have been stained loose and have been taken to 95% alcohol:

- Place sections in a mixture of equal parts of xylene, chloroform and absolute alcohol for not longer than 5 minutes.
- Transfer sections to a dish of pure beechwood creosote for not longer than 7 minutes.

- Float the section onto a slide. The method is as for paraffin sections, but celloidin sections are more robust and can be picked up with a pair of forceps.
- Blot with hard toilet paper (the sort that comes in boxes rather than as rolls; do not use soft toilet paper. The older books recommend the use of cigarette paper: more available in laboratories in days gone by than now).
- Pour xylene over the slide, drain and repeat. Do this quickly as the xylene may cause the section to crease. Curling of the sections at this stage can be a problem.
- Blot rapidly with the hard toilet paper.
- Mount in DPX. Use plenty of DPX: it will be absorbed by the celloidin and tends to retract under the coverslip. Davenport's great experience of handling celloidin sections is worth drawing on. He recommended weighting the coverslips with 1.5–2.0 oz lead weights. A 50 g weight would be suitable.

CHAPTER 7

Standard Staining Techniques

Staining of histological sections is both an art and a science: neither aspect should be ignored. In this chapter the technical aspects of staining are considered; the scientific principles underlying staining are considered in a later chapter. We begin by describing the rather simple apparatus needed for staining small batches of sections. In most laboratories staining is often done using staining machines and these, when properly adjusted and used, can produce excellent results. The beginner should always start by staining his or her own material by hand. A series of standard techniques have been set out step by step. Attention has been paid to the use of the staining microscope. This allows the depth of stain to be judged and, as necessary, adjusted. None of the techniques described here are difficult, though some are lengthy and fiddly to get right. Practice is, of course, the solution to most problems that the beginner is likely to encounter. The haematoxylin and eosin (H&E) method has been considered in some detail. This is the most widely used of all histological staining methods and will almost certainly be used by the reader. Familiarity with the H&E method sometimes breeds contempt and not enough attention is paid to producing first-class H&E stained sections. The beginner is encouraged to practise his or her H&E staining until the procedure become automatic and the results uniformly excellent.

Histological Techniques: An Introduction for Beginners in Toxicology
By Robert Maynard, Noel Downes and Brenda Finney
© Maynard, Downes and Finney 2020
Published by the Royal Society of Chemistry, www.rsc.org

7.1 INTRODUCTION

A justifiable sense of achievement or relief may be felt when you have a series of sections ready for staining. Staining is the part of section preparation that is the most fun and certainly the most interesting. In this chapter the apparatus needed for staining is described and then a limited number of standard methods are set out in some detail. The number of methods has been deliberately restricted: there are enough for most studies where the histopathological effects of chemicals are being investigated. Many more techniques may be found in the works listed in the Bibliography.

Staining combines art with science. Of course, our interest is in the science, but it is a poor histologist who cannot recognise the beauty of a perfectly stained histological section. Staining usually involves bringing out the various features of a section in different colours. The colours themselves are not important, but the contrast between the colours is. Thus, nobody would stain both the nuclei and the cytoplasm reddish-pink, though this would be easy enough to do with Nuclear Fast Red and eosin. Keeping the intensity of the colours at a correct level requires practice: over-staining is perhaps a greater fault amongst beginners than under-staining.

Staining paraffin wax sections involves removing the wax from the section and then placing the section in a medium that conforms to that of the stain. Wax is removed by placing the section in xylene for a couple of minutes. The time is not critical. Warming the section to just melt the wax used to be advised because it speeded up the removal of the wax by the xylene. This is not necessary unless you are in a great hurry. After the xylene, place the section in absolute alcohol (1 minute), 90% alcohol (1 minute), 70% alcohol (1 minute) and then distilled water. This "running down the alcohols" sequence will be reversed when the section has been stained and we are taking the section back to xylene before mounting with a medium, such as distrene, plasticiser, xylene (DPX), which is miscible with xylene. Some experts advise a longer series of alcohols (50%, 30%), but this is not necessary. Why not go from absolute alcohol to water in one jump? This would set up diffusion currents that might pull the section off the slide: at least, that is what is usually said. I doubt it makes much difference but sticking to a routine that was established more than one hundred years ago is probably wise.

7.2 APPARATUS NEEDED FOR STAINING

Keeping a couple of feet of bench space for staining is a good idea. The surface of the bench should be covered with a disposal cover or with a sheet of toughened plate glass placed over a sheet of white paper. A sink is essential. Find a place for the staining microscope: to one side of the staining area. Some shelves directly in front of you are useful: staining involves the use of certain solutions and organic solvents again and again: having them within easy reach makes good sense.

Staining involves immersion of the slide carrying the tissue section in stain. This can be done in several ways. Special glass or plastic jars with slotted interiors (Coplin jars, see Figure 7.1 top panel) are very useful if only a few slides are to be stained. For larger numbers, racks and staining troughs are better. The metal racks take 20 or so slides and can be easily transferred from trough to trough (Figure 7.1 bottom panel). Staining can also be done by placing the sections on glass rods over the sink and pouring on the stain. The glass rods should be anchored at a suitable distance apart: metal clamps, Plasticine® or drilled wooden blocks accomplish this. Staining on glass rods is messy: the sink usually ends up in a terrible state. Staining in jars is recommended. Old fashioned glass Coplin jars have one fault: the glass lids fit only loosely and volatile liquids evaporate. The glass staining troughs also tend to have loosely fitting lids: plastic jars with screw tops and troughs are better in this respect.

Those who have just begun staining can be recognised by the rainbow appearance of their hands. Slides have to be moved from jar to jar and the fingers get into the stain, at least this was my experience. Dilute bleach removes most stains from the skin but be careful: it is much better not to get the stain on your fingers. Gloves? Yes, or use forceps to move the slides from jar to jar. Special forceps for gripping glass slides are available: they have broad flat ends. I found that an ordinary pair of blunt dissecting forceps was perfectly satisfactory. They can be improved by pushing some narrow bore rubber tubing over the tips of the forceps: this softens the grip on the slides and prevents slipping.

Slides will need to be washed. Washing with distilled water is done by means of a wash bottle: the effluent runs into the sink. Washing with tap water (usually for longer than washing with distilled water) is best accomplished in a staining trough into which water is being run *via* a narrow rubber tube. Put the trough in the sink.

A stop clock is needed for timing stains.

Slides can also be stained, at least with the simpler stains, by means of an automatic staining machine (Figure 7.2). Very high-quality haematoxylin and eosin (H&E) stains can be produced in this

Figure 7.1 **Hand-staining set-ups.** In the top panel we show a Coplin Jar, which are ideal if you are just staining a single slide or a handful of slides. The bottom panel shows several troughs for processing larger numbers of slides. You may not get the consistency of an automated stainer with a set-up like this, but good batch results can be produced in a reasonable amount of time.

way. But more complicated stains require finer control, and differentiation under the staining microscope is essential. Automatic staining machines are of great help in laboratories processing large numbers of sections, but the beginner would be well advised to learn how to stain by hand before going on to use such machines.

Figure 7.2   **Automated stainer.** Great if you have access to one, but an expensive item. If you have one hundred or more slides to do, it may be worth approaching one of the contract labs who will stain them for you at a price.

7.3   STANDARD METHODS

The techniques described here may be applied from paraffin wax sections to celloidin sections to frozen sections. Because paraffin wax sections are most widely used, the methods will be described for this type of section. At the end of the chapter, some notes on the staining of celloidin and frozen sections are provided.

7.3.1   Haematoxylin and Eosin, Iron Haematoxylin Stains

Haematoxylin and eosin, H&E, is the most widely used combination of all stains available to the histologist. It provides good differentiation of nuclei and cytoplasm and, as an oversight stain, is so well established that criticism is perhaps pointless. Given that H&E is so widely used, it seems appropriate to provide a fairly detailed description of the stain and how the combination is, or should be, used.

7.3.1.1 Haematoxylin. The formula of haematoxylin is shown in Scheme 7.1 alongside that of haematein.

   It should be known that haematoxylin, a colourless compound, must be oxidised to haematein to produce the red (hence the name, haem, from the Greek *haimato,* meaning blood, and xylon, meaning wood) dye used by histologists. Baker (1958) provided an interesting account of the origin and history of haematoxylin. Haematoxylin is

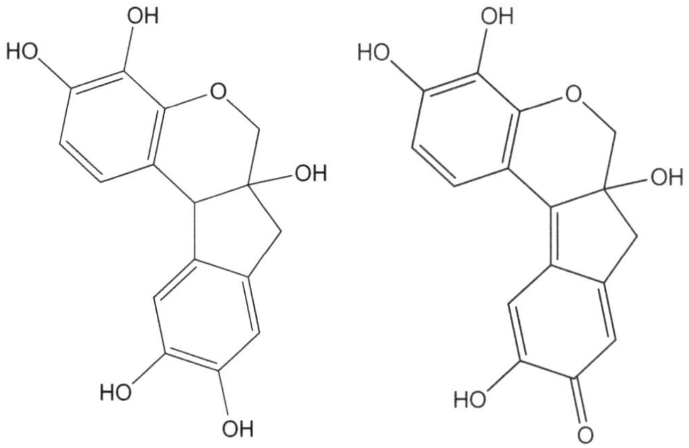

Scheme 7.1 The chemical structures of haematoxylin (left) and haematein (right).

extracted from the heartwood of the tree *Haematoxylon campechia-num*, a smallish leguminous tree found by the Spanish in Mexico and now grown in the Caribbean. The heartwood is exported as "logwood chips": a name familiar to all those who, as children, were given a "chemistry set" as a Christmas present. The dye has been used for dying wool, silk, leather and nylon. It was introduced into histology in the 1840s and various mordants were then explored by Boehmer in 1865 (Mann, 1902). Conn pointed out that Hooke had used "tinctures of logwood and cochineal" to stain hairs in 1665; Conn added that whether or not the hairs were coloured is unknown.

Haematein is a reddish dye but makes a poor stain unless a mordant is used to link the stain to tissue. In combination with a suitable mordant, haematein becomes an excellent dye. The presence of a mordant also changes the colour of the dye form reddish-brown to blue, purple or black. The mordant used in histological methods involving haematoxylin (actually haematein) are alums. "Alum" is derived from the French *alute* meaning tawing. Tawing is a process akin to tanning and is, or was, used in the preparation of leather: alum was used in the process. Alums are double salts: aluminium is often one of the cations present, but ferric alum contains iron and chrome alum contains chromium. The second cation is usually potassium, ammonium or sodium. All alums crystallise with 24 molecules of water in the alum molecule. The word "aluminium" is derived from the word "alum" and not *vice versa*.

Two reactions are thus necessary to convert haematoxylin into a useful histological dye. First, the haematoxylin must be oxidised to haematein and, second, alum must be present to allow binding to tissue. Haematoxylin can be dissolved in alcohol (ethyl alcohol is generally used) and allowed to stand in sunlight. Oxygen in the air oxidises the haematoxylin to haematein. Alternatively, an oxidant can be added to the solution to produce immediate oxidation. Many different oxidants have been used: sodium iodate and mercuric oxide are often used. The process of oxidation is referred to as ripening of the stain. The alum may be added before or after ripening. Interestingly, there seems to be no specific reason for using alums as mordants. The metallic cation seems to be the essential component and aluminium sulphate works as well as potassium alum, which contains aluminium. Ferric chloride can also be used.

Many formulae or recipes for alum haematoxylin stains have been developed. It is usual to distinguish between the "alum haematoxylins" that contain potassium or ammonium alum and those that contain iron alum: the "iron haematoxylins". The latter include Heidenhain's and Weigert's haematoxylins; the former include, of many, Ehrlich's, Mayer's, Delafield's and Harris' haematoxylin. Iron haematoxylins stain nuclei deep brown or black; the alum haematoxylins stain nuclei blue–purple. An important difference between the two groups lies in the ease with which the stain can be removed from nuclei by acid. The brown–black staining produced by the iron haematoxylins is difficult to remove, the blue–purple stain produced by the alum haematoxylins is not. Another important difference is that the alum haematoxylins initially stain nuclei deep red and this colour is converted to blue–purple by exposing the sections to an alkaline wash. This process is referred to as bluing the sections. Bluing is not necessary with the iron haematoxylins. Slight differences in the shade of blue–purple are produced by the different recipes for alum haematoxylins: Ehrlich's produces rather bright blue–purple nuclei; Harris' produces a deeper purple shade.

All haematoxylins are basic stains. They bind to acidic components of cells. In the nuclei the acidic components are provided by nucleic acids; in cytoplasm the RNA in ribosomes, for example, provides the acidic components. Because of the large amount of nucleic acid in the nuclei, the haematoxylins are effective nuclear stains. Basic stains are more highly ionised at low pH values; adding acid to the stain sharpens the staining of nuclei. Lowering the pH also reduces the staining of non-acidic components of cells. All components of cells will, eventually, take up the haematoxylin stain: both nuclei and cytoplasm will be stained blue or black after some time. The nuclei take up the stain more rapidly than the cytoplasm and this allows for what is called progressive staining. Sections are exposed to the stain until the nuclei are well stained but not much stain has been taken up

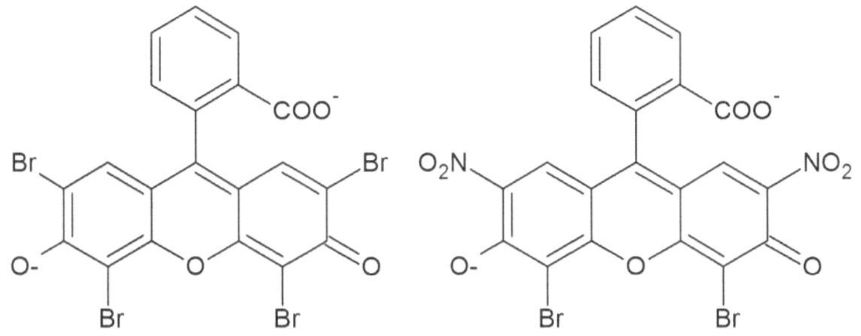

Scheme 7.2 The chemical structure of eosin Y (left) and eosin B (right).

by the cytoplasm, and then sections are removed from the stain and blued in tap water. The alternative method is described as regressive staining: sections are allowed to over-stain in the haematoxylin and then exposed to dilute acid (acid alcohol: 1% hydrochloric acid in 70% ethyl alcohol). The acid rapidly removes the dye from the cytoplasm but more slowly from the nuclei. The process will be explained when the use of Ehrlich's haematoxylin is described. Progressive staining is inevitably less precise than regressive staining in that no visual control of the stain is involved.

Having stained nuclei with haematoxylin, it is conventional to stain the cytoplasm with an acid stain. A wide variety of dyes may be used: they are sometimes referred to as plasma stains. Eosin is easily the most widely used of such stains.

7.3.1.2 Eosin. Eosin is a pink–red dye of the xanthine series. Its formula is shown in Scheme 7.2. A number of different eosins are available. Eosin Y (Y standing for yellow, CI: 45380) produces a slightly yellowish pink stain; eosin B (B standing for blue, CI: 45400) produces a deeper red (not blue!) stain. Eosin Y is the most widely used form. The word eosin is derived from the Greek *eos*, meaning the dawn. Eosin Y, when used properly, produces a pale yellowish pink, very like the colour of the dawn sky. This should be remembered: the flaming red–pink colours (sometimes described as sunsets) produced when eosin is badly used is an error of technique and reduces the value of the stain. This is discussed in more detail below.

7.3.2 The Staining Microscope

The staining microscope is mentioned, often, in the following sections: it might be as well to describe it now. Examination of sections taken from staining solutions is often necessary in order to control the depth of staining and to allow differentiation of the staining of

Figure 7.3   **A full set of tissues stained with H&E using an automated staining machine.**

different components of the section. The section will be examined without a coverslip and stain will get onto the stage of the microscope. Examination at low and medium power will be all that will be needed. It is sensible to keep an old microscope for this purpose. Always clean the stage after use. I found that the mechanical stage fitted to all serious microscopes was a nuisance when staining and usually removed it. An old monocular microscope, using daylight illumination or illumination from a lamp, is perfectly satisfactory. Regular use of the staining microscope is advised. Figure 7.3 shows the end result of H&E staining produced on a set of tissues by an automated stainer.

7.3.3   Technique for Staining with Alum Haematoxylin and Eosin

7.3.3.1   Ehrlich's Haematoxylin and Eosin.   This is a widely used stain and produces excellent staining of nuclei.

Recipe

The haematoxylin solution

Haematoxylin	2 g
Absolute alcohol	100 mL
Glycerol	100 mL
Distilled water	100 mL
Glacial acetic acid	10 mL
Potassium alum	10–14 g (in excess)

The haematoxylin is dissolved in the alcohol and the water and the other components then added. The solution should be placed in a wide-necked flask and placed in the sun, for example, on a window-sill. Now wait. Ripening takes some weeks. As the stain ripens, it changes in colour from purple to deep red and it develops a splendid vinous aroma, rather like that of port wine. Gently shake the flask occasionally. The glycerol is included to retard evaporation; the acetic acid sharpens the stain. How will we know that the stain is ready for use? Only by testing it on sections. Once ripened, the stain will keep for some months but it will, eventually, deteriorate. This is due to over-oxidation. When testing the stain, choose a tissue with a lot of nuclei: lymph node is an excellent choice. Ehrlich's haematoxylin, like most other stains, can be bought "ready made".

Method

De-wax and take sections to water. This standard phrase means running the section through xylene and a series of baths of descending concentrations of alcohol. The series is reversed when the section is dehydrated and cleared prior to mounting. The timings are not critical:

Xylene	2 minutes
Absolute alcohol	1 minute
90% alcohol	1 minute
70% alcohol	1 minute
Distilled water	Rinse well and the sections are ready for staining in an aqueous solution of dye.

If the tissue has been fixed with a mercury-containing fixative, re-move the mercury pigment as described in the chapter on fixatives (Chapter 5).

Stain in a Coplin jar of Ehrlich's haematoxylin for 10 minutes. A shorter time may be adequate but over-staining does not matter because we will be differentiating the stain in a moment.

Rinse in water to remove excess stain. Rinse by dipping a few times in a beaker of water or with a slow-running stream of water from the tap.

Wash in tap water for 5 minutes. This is best done by putting the slide or slides in a rack, immersing the rack in water in a glass trough and running water, gently, into the trough *via* a narrow rubber tube connected to the tap. This is the bluing process. The effectiveness of the process depends on the pH of the tap water. In areas with very soft water (acid water) the process will not work and a substitute for tap water must be used. Scott's tap water substitute is excellent: for the formula, see the Appendix. Bluing in the tap water substitute takes only 15 or so seconds.

Examine the section under the staining microscope. Use a 10× or 25× objective and look closely at the nuclei. These should be blue–purple in colour. The whole section, in fact, will probably be blue–purple in colour but the nuclei will stand out clearly.

Dip the section in 1% acid alcohol (a mixture of 1 mL concentrated HCl and 99 mL 70% ethyl alcohol).

Rinse and blue again in tap water.

Examine with the staining microscope.

The cytoplasm of the cells, but not the nuclei, will have lost colour.

Repeat the destaining and bluing until only the nuclei are coloured. If you overshoot and remove the colour from the nuclei, you can simply wash the section in water and put it back in the stain. Then repeat the differentiation. We are using the haematoxylin as a regressive stain: we over-stain and then differentiate. It can be used as a progressive stain: simply stain in the haematoxylin and then in the eosin as described below. This is less satisfactory: controlling the extent of staining with the haematoxylin is the key to producing really good preparations.

This seems a slow and complicated process and so it is at first. However, once the strength of the haematoxylin has been gauged, the time needed in the acid alcohol and in the bluing bath can be standardised. In practice, a quick rinse in the acid alcohol and a few minutes of bluing is all that is needed. Removal of every trace of acid used for differentiation is important: if this is not done then the nuclear stain may fade. Some workers have suggested washing in 1% sodium bicarbonate to ensure that absolutely all the acid had been neutralised. In general this does not seem to be necessary, but wash well with tap water before staining with eosin.

Once the nuclear staining is complete, transfer the sections to a jar of 1% eosin (1% in distilled water) and stain for 1 minute.

Rinse in tap water and examine under the staining microscope.

The nuclei will be stained blue, the cytoplasm will be pink.

Dehydrate rapidly through the alcohols: a dip in the 70%, 30 seconds in the 90% and 1 minute in the absolute alcohol.

Dip in xylene for 1 minute.

Mount in DPX.

Allow the DPX to set until it is hard, clean up the slide and examine.

Examination of the finished section

Almost inevitably, you will be pleased: the nuclei will be blue and the background pink. You may now wish to look a little more critically. Examine the nuclei under high power: note that the chromatin has stained deep blue. Note that the nuclear membrane is well defined and that the nucleolus, if you can see it, is a pale pink. If you are examining a section of kidney tubule, you might look towards the base of the cell in the hope of identifying mitochondria. These will appear as a series of very fine blue striations. Mitochondria are effectively destroyed by some fixatives: see Chapter 5. Now look for collagen fibres. These will not stand out, but will be stained pink, rather like the cytoplasm of the cells. Some variation in the depth of staining may be visible, but this is unlikely to be marked. The deeper the eosin staining, the less differentiation there will be between cytoplasm and extracellular components. If your section seems to have been over-stained with eosin, look back to the method and reduce the time in the eosin bath, or allow a little longer in the 70% alcohol. Under-staining with eosin is less of a problem than over-staining: remember the pale pink of the dawn sky!

What makes a really good H&E stain? (See Fig. 7.3 and the figures in previous chapters for good examples.)

1. Perfectly stained nuclei with chromatin bright blue–purple and fine staining of thin threads of chromatin material. The nuclear membrane should be clear. The nuclear background (the regions between the chromatin) should be pale blue or blue–pink.
2. Pink staining of nucleoli or at least of parts of the nucleoli.
3. Translucent pale pink staining of cytoplasm, with the shade of the stain varying with the amount of RNA present. In cells containing a lot of RNA, for example, plasma cells, the cytoplasm will stain more bluish than pink.
4. Keratin should be stained orange–pink and should be distinguishable by the shade of the stain from cytoplasm.
5. Muscle fibres and elastic fibres: deep pink, a deeper pink than the cytoplasm of epithelial cells.
6. Collagen: light pink and distinguishable from muscle.
7. Elastic fibres: a glossy pink. This is difficult to describe but you will see what I mean if you stain a section of an artery and look at the internal elastic lamina.
8. Cartilage: pink–light blue. This is rather variable.
9. Mucin: light blue or blue–pink.

You will see that much depends on shades of pink! A few points relating to the staining with eosin are worth mentioning. Formalin tends to inhibit staining with eosin: a little longer in the eosin bath is needed after fixation with formalin than after fixation with a solution containing mercuric chloride. If you have difficulty in keeping the eosin in the section, you can add eosin to the dehydrating baths. Eosin is less soluble in alcohol than in water and, once you have got past the 90% alcohol, little more eosin is likely to be extracted from the section. Some histologists prefer a 0.5% eosin solution to a 1% solution: staining takes a little longer but better differentiation of the pink-staining components of the section may be produced.

7.3.4 An Alternative Haematoxylin: One of Many

7.3.4.1 Harris' Haematoxylin

Recipe

Haematoxylin	1 g
Absolute alcohol	10 mL
Ammonium or potassium alum	20 g
Distilled water	200 mL
Mercuric oxide	0.5 g

Method

Dissolve the alum in hot distilled water. Dissolve the haematoxylin in the alcohol and add the alum solution. Boil. Add the mercuric oxide: the purple mixture will froth vigorously so use a large flask! Cool under the cold tap. Filter before use.

The procedure is identical with that for Ehrlich's haematoxylin but differentiation is obligatory: this is a regressive rather than a progressive stain (see section 7.3.1.1 and Chapter 8). Nuclei will be stained a darker purple than with Ehrlich's haematoxylin.

7.3.5 Substitutes for Eosin

Given that eosin is so widely used, it might be interesting to know that there are other plasma stains that are at least as good, and perhaps better, than eosin. Biebrich scarlet (CI: 26905, 1% in distilled water), Chromotrope 2R (CI: 16570, 1% in distilled water) and Erythrosin B (CI: 45430, 1% in distilled water) are all very good counter-stains for haematoxylin and are less likely to be removed from the section during dehydration.

7.3.6 Another Alternative Haematoxylin Stain

J R Baker introduced a staining method referred to as "Haematal 8 and Biebrich scarlet". This is an outstanding stain and no differentiation step is required.

Recipe

Solution A: aluminium sulphate (hydrated: $16H_2O$): 15.76 g; distilled water: 1000 mL

Solution B: Haematein (Note: **Not** haematoxylin!): 1.876 g; ethylene glycol, 50% in water: 1000 mL

Mix equal volumes of solutions A and B just before use. The mixture does not keep well. The rather over-exact weights can be rounded to one decimal place.

Method

De-wax and take sections to water; remove mercury pigment if present.

Haematal 8	10 minutes
Wash in tap water	3 minutes
Rinse in distilled water	
Biebrich scarlet	30 seconds

Rinse in distilled water

Dehydrate, dip in xylene and mount in DPX.

This stain produces good differentiation of cytoplasm and extracellular components of the section. Bradbury recommended this stain.

7.3.7 Criticisms of H&E Staining

Can so widely used a stain be criticised? After all, much of what is known of pathological processes has been learnt from sections stained with H&E! The answer to the question is certainly yes. Gabe, a most distinguished French histologist, found no place for a detailed description of the alum haematoxylin and eosin method in his exhaustive 1106 pages; Lillie (1976) (see Bibliography) provided a detailed account but added, "Excellent as the haematoxylin–eosin stains are for permanence, we prefer for routine use procedures of the azure–eosin type." It is, of course, important not to mix up criticisms of the stain with criticisms of how well it is used. All stains can be used badly and all can produce excellent results in the hands of an expert. Carleton, a master of histological technique, had no reservations about haematoxylin and eosin staining and recommended it strongly.

For the toxicologist interested in histology, the major criticism of H&E staining is its failure to differentiate between normal cells and cells undergoing the early stages of necrosis. This makes looking for early signs of cellular damage rather difficult. This problem can be solved by use of Lillie's azure A and eosin B method. There is a strong case for using this as the routine stain in toxicological pathology practice.

7.3.7.1 Lillie's Azure A and Eosin B Method

Recipe

Azure A	4 mL of a 0.1 % solution in distilled water
Eosin B	4 mL of a 0.1% solution in distilled water
0.2 M acetic acid	0.7 mL
0.2 M sodium acetate	1.3 mL
Pure ("AnalaR") acetone	5 mL
Distilled water	25 mL

Discard the solution after use: it does not keep well.

The acetic acid/acetate buffer is critically important: the proportions given above produce a pH of 5–5.5 and this has been shown to be ideal for staining tissue fixed in solutions containing mercuric chloride. If formalin has been used then the proportions should be 1.7 mL 0.2 M acetic acid/0.3 mL 0.2 M sodium acetate, producing a pH of about 4. Other buffers might be used but sticking to Lillie's recommendation seems to be sensible.

Method

De-wax and bring sections to water, removing mercury pigment if necessary.

Stain for 60 minutes in the Azure A-eosin B mixture.

Acetone	1 minute, 3×
50/50 acetone and xylene	1 minute
Xylene	1 minute, 2×
Mount in DPX.	

Note that no differentiating step is needed, this is an automatic stain and can be used with staining machines.

Results

The distribution of the blue and pink shades is controlled by the pH of the staining solution. Increasing the pH will increase the number of structures that stain blue, decreasing the pH will tilt the balance towards pink staining. Before using the stain as a research tool, it would be sensible to experiment with the effects of changing the pH of the staining solution. The critical point is that normal cytoplasm should stain a pale blue–lavender–violet and the cytoplasm of damaged cells should stain pink. This differentiation will not be produced if an inappropriate pH is used. Lillie provided a long list of staining characteristics of various components of tissue:

Nuclei	Blue
Mast cell granules	Blue–violet
Granules of basophil leucocytes	Blue–violet
Cartilage matrix	Reddish-violet
Calcium deposits	Dark blue
Cytoplasm of normal cells	Light blue–lavender–violet
Cytoplasm of necrotic cells	Bright pink
Hyaline degeneration products	Pink
Red blood cells	Orange–red
Mucin	Pale greenish to blue to blue–violet
Plasma cell and lymphocyte cytoplasm	Blue
Secretory granules of salivary glands, pancreas and Paneth cells of the gut	Red–pink
Collagen	Pink or blue depending on the pH of the stain.

It will be appreciated that the range of colours is much wider than that produced by the H&E method. Gabe warmly recommended Lillie's azure A–eosin B method.

7.3.8 Iron Haematoxylin

Heidenhain's iron haematoxylin is the finest of all iron haematoxylin stains and was the stain most used by cytologists in the pre-electron microscopy period. It remains a splendid method but is a slow one and is, perhaps, rather tedious to use. I did not find it to be so; the reader is referred to the Bibliography for accounts of this splendid stain. Weigert's iron haematoxylin is an easier and quicker method and is entirely satisfactory for routine histological work.

7.3.8.1 Weigert's Iron Haematoxylin. Weigert's iron haematoxylin stain is used when staining for collagen fibres with the van Gieson method: it resists removal from the nuclei very much better than the alum haematoxylin stains.

Recipe

Two solutions are needed:

A	Haematein	1 g
	Absolute alcohol	100 mL
B	30% solution of ferric chloride in distilled water	4 mL
	Distilled water	100 mL
	Concentrated hydrochloric acid	1 mL

Solution A and B keep well in stoppered bottles.

Method

Take sections to water in the usual way, removing mercury pigment if necessary.

Stain for 20 minutes in a 50/50 mixture of solutions A and B. The mixture does not keep well: make up when you need it.

Rinse in tap water.

Differentiate the nuclei (see note below) with acid alcohol until only the nuclei are stained dark brown or black.

Wash in running tap water for 5 minutes.

Counter stain, if desired, with eosin (see section 7.3.3.1).

Dehydrate, clear and mount.

The nuclei are stained a deep blue–black. The differentiation step is not needed if Wiegert's iron haematoxylin is being combined with the van Gieson method for collagen: the picric acid in the van Gieson stain removes quite enough of the iron haematoxylin.

7.3.9 The Celestine Blue-haemalum Method for Staining Nuclei

This method was introduced as an alternative to the iron haematoxylin methods. It may be used whenever an acid-resistant nuclear stain is required.

Recipe

Solutions

Celestine blue

Celestine blue	0.5 g
Ferric alum	5 g
Glycerol	14 mL
Distilled water	100 mL

Dissolve the ferric alum in warm distilled water, add the Celestine blue and bring to the boil for 3 minutes. Cool and add the glycerol.

Mayer's haemalum

Haematoxylin	1 g
Sodium iodate	0.2 g
Potassium alum	50 g
Citric acid	1 g
Chloral hydrate	50 g
Distilled water	1000 mL

Dissolve the haematoxylin, sodium iodate and potassium alum overnight in the distilled water. Add the chloral hydrate and citric acid and boil for 5 minutes. Cool.

The chloral hydrate is said to act as a preservative and the citric acid is said to sharpen the stain.

Method

De-wax sections and bring to water, remove mercury pigment as necessary.

Celestine blue	5 minutes
Rinse in tap water.	
Mayer's haemalum	5 minutes

Wash well in running tap water.

Differentiate in 0.25% (or 0.5%) HCl in 70% alcohol. The usual 1% acid alcohol could be used but the more dilute solutions allow better control.

Wash well in tap water.

If the stain is being used as a prelude to some other method, proceed as specified in that method. If the stain is being used alone then counter-staining with Orange G (0.5% in 90% alcohol) is recommended. Eosin or any other plasma stain could be used.

Results

Nuclei	Black
Cytoplasm	According to the counter-stain

7.4 CONNECTIVE TISSUE STAINS

The topography of tissues is more easily studied if connective tissue is stained. In addition, there are good reasons for staining connective tissue itself and for examining its response to injury. Wound healing involves the deposition of reticulin fibres and, later, collagen fibres. Fibrosis, a common response to injury by chemicals and other agents, involves the deposition of collagen fibres. Elastic fibres may be damaged, as in emphysema, and damage to the elastic laminae of arteries is seen in atherosclerosis. Basement membrane contains reticulin fibres and is thickened in diseases of the airways, especially in asthma, and in diseases of the kidney. Fibrin, though not a connective tissue, will also be considered in this section: fibrin appears wherever plasma escapes from vessels, as in the inflammatory response and, of course, when blood clotting occurs.

It is possible to stain all the elements of connective tissue in a single section, but the methods developed to allow this are, inevitably, complicated and, unless great care is taken, rather messy preparations can be produced; much better, perhaps, is to stain serial sections for each component of connective tissue. A very large number of connective tissue stains have been developed and manuals of histological techniques contain dozens of variants on many of the basic staining types. Here, we shall discuss only a small number of stains that have been shown to be reliable. Even some of these are fairly complicated.

7.4.1 Staining of Collagen Fibres

There is little difficulty in staining collagen fibres; staining **all** collagen fibres, including those recently laid down, is, however, much less easy. Distinguishing collagen fibres from reticulin fibres requires special stains for each fibre type. Elastic fibres, containing elastin, are not difficult to stain. Many different types of collagen have been identified. Type I is very widely distributed, type II occurs mainly in cartilage, type III appears early during wound healing and as reticulin fibres and type IV collagen occurs in basement membrane. The other types of collagen require specific antibody techniques for their identification and occur in much smaller amounts than the main types listed here.

7.4.1.1 Standard Collagen Stains. Three methods will be described: van Gieson's, Masson's trichrome and Lillie's HCl–Biebrich scarlet–methyl blue. Figure 7.4 shows the van Gieson and Masson methods and the MSB method, see 7.4.2.1. Heidenhain's azan stain (azocarmine and aniline blue) is perhaps better than any of these but is a complicated method and takes longer than the methods described here: see handbooks in the Bibliography for details of the azan method. Gabe recommended the azan method strongly; Lillie regarded it as too "cumbersome" for routine use.

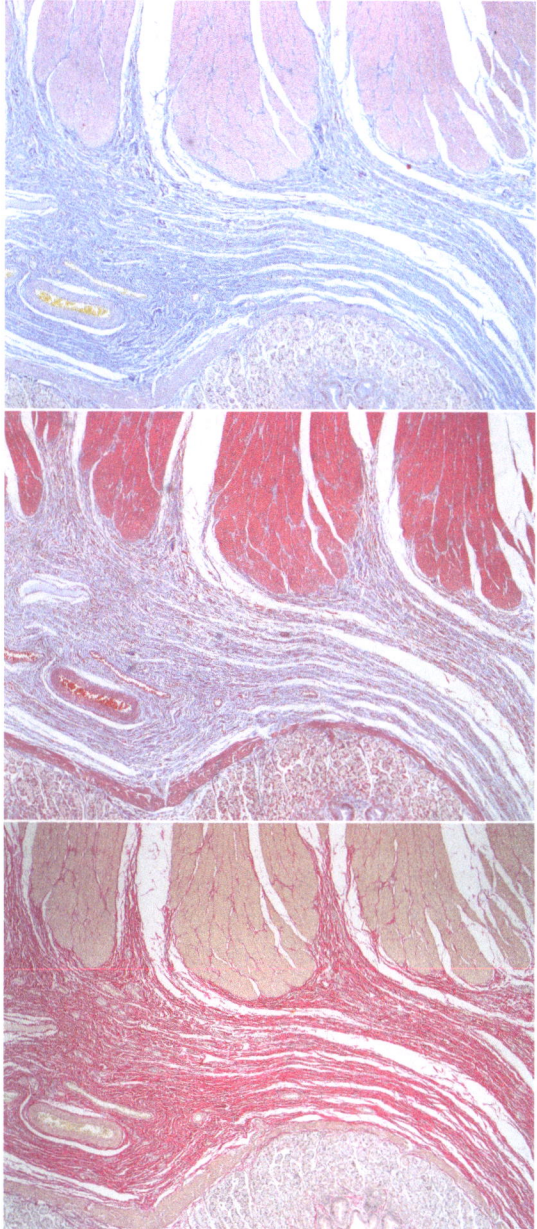

Figure 7.4 **Three connective tissue stains.** Top: MSB, middle: Masson's, bottom: Van Gieson's. All stain collagen but only the MSB method distinguishes fibrin.

The van Gieson picro-acid fuchsin method for collagen

(Note: fuchsin might be more accurately written as fuchsine.)

This is a very widely used and reliable method for staining collagen fibres. It is fair to say that it is less effective in demonstrating recently deposited collagen than "old" collagen. Although widely used, it is often not well used and the differentiation of collagen fibres from, for example, muscle is often not as sharp as might be wished. The objective should be to produce a section with black nuclei, pale lemon-coloured muscle and bright red collagen fibres. Bleeding, for want of a better word, of the red stain into the yellow is often seen and muscle is stained an orange–brown. A number of variants of the basic method have been introduced in an attempt to sharpen the differentiation. The method recommended here was introduced by Lillie in 1945.

Recipe

Solutions

Acid fuchsin	100 mg in 100 mL distilled water, to which 0.25 mL conc. HCl is added
Picric acid	Saturated aqueous solution

Mix 5 mL acid fuchsin with 95 mL picric acid to produce the staining solution.

Method

De-wax sections and take to water, removing mercury pigment if a mercuric chloride fixative has been used.

Stain in Weigert's haematoxylin (see section 7.3.8.1) for 15 minutes.
Wash in running tap water for 5 minutes.
Rinse in distilled water.
Stain in the picric acid–acid fuchsin mixture for 5 minutes.
Rinse in distilled water and then:

95% alcohol	1 minute; 2×
Absolute alcohol	1 minute; 2×
Xylene	1 minute
Mount in DPX.	

Results

Collagen	Bright red
Muscle and cytoplasm	Yellow (generally)
Nuclei	Black

Points to note

The iron haematoxylin is removed from all but the nuclei by the picric acid: differentiation is needed.

The collagen stain is removed by alkaline tap water, hence the dehydration in 95% alcohol.

Pausing to look at the sections during the dehydration in 95% alcohol is useful: use the staining microscope. Washing in the alcohol will remove the yellow stain of picric acid: the alcohol acts as a weak acid-differentiator. It is likely that you will not need to reduce the picric acid staining. Reducing the red collagen staining is easily accomplished: dip in tap water, rinse in 95% alcohol and examine again.

If you have difficulty in keeping the yellow of the picric acid in the section then adding alcoholic picric acid to the absolute alcohol bath is helpful. Picric acid is supplied under water, it is explosive when dry. To make an alcoholic solution: add water-moist picric acid to absolute alcohol until a layer of undissolved picric acid lies at the bottom of the vessel. Decant the solution and add sufficient absolute alcohol to dissolve most of the solid material. Decant the saturated alcoholic solution. Using this instead of absolute alcohol in the dehydration sequence will ensure bright yellow staining. Dehydrating in acetone also helps to preserve the yellow stain.

Better colours are produced after fixation in fluids that contain mercuric chloride.

Timings are very dependent on the tissue, the fixative and the sample of acid fuchsin used. The latter is unlikely to be a problem. Varying the timings until you produce optimal staining of your material is necessary.

Masson's trichrome stain for collagen

In this method the nuclei are stained with iron haematoxylin and the collagen with either a blue or green stain (aniline blue or light green or fast green) and the cytoplasmic background with acid fuchsin and Ponceau 2R. Phosphomolybdic acid is used to prevent staining of collagen with acid fuchsin and to reserve it, so to speak, for staining with the blue or green dye. Phosphomolybdic acid is sometimes described as a mordant but this is quite incorrect: see Chapter 8. Whether a blue or a green collagen stain is used is a matter of taste. Well-stained sections show excellent colour contrast. Over-staining tends to produce a muddy mixture of colours. The aim should be for light but very sharply differentiated staining.

Recipe

Solutions

Plasma stain

Ponceau de xylidine (Ponceau 2R, CI: 16150)	1% in 1% acetic acid (50 mL), mixed with:
Acid fuchsin (CI: 42685)	1% in 1% acetic acid (25 mL)

Phosphomolybdic acid
1% in distilled water

Collagen stain (choose **one** of the following)

Aniline blue (CI: 42755)	2% in 2% acetic acid
OR	
Light green (CI: 42095)	2% in 1% acetic acid
OR	
Fast green (CI: 42053)	2% in 1% acetic acid

Acid wash
Acetic acid (1%)

Method

De-wax and take sections to water. If the tissue has been fixed in formaldehyde then soaking the sections in saturated aqueous picric acid containing 3% mercuric chloride for 3 h followed by washing in tap water and removal of mercury pigment is recommended. This helps the dying of the collagen fibres.

Stain with Weigert's haematoxylin: see section 7.3.8.1.

Wash well in running tap water for 5 minutes.

Differentiate in 0.5% HCl in 70% alcohol until only the nuclei are stained. The usual 1% acid alcohol may be used but the 0.5% solution allows better control. Haematoxylin staining of the cytoplasm and other elements of the tissue will obscure the later stains.

Wash well in running tap water for 5 minutes.

Rinse in distilled water.

Stain in the plasma stain for 5–10 minutes.

Rinse in distilled water.

Differentiate in phosphomolybdic acid. Continue until the collagen fibres are decolourised.

Rinse in distilled water.

Stain in the blue or green collagen stain for 2–5 minutes.

Differentiate in 1% acetic acid until only the collagen is stained green.

Dehydrate, dip in xylene and mount in DPX.

Results

Nuclei	Black
Muscle, red blood cells, fibrin, some cytoplasmic granules	Red
Collagen, some reticulin, amyloid, mucin	Blue or green according to stain

The staining should be light but perfectly sharp. Sections that appear dark green or blue with a hint of murky red and muddy brown–black nuclei are unsatisfactory. Sections of lung provide excellent test material. Examine the wall of a large bronchus: the smooth muscle, the collagen and the mucin of the goblet cells and submucosal glands should all be clearly differentiated.

Points to note

The time needed in the plasma stain varies with the fixative used. If the tissue has been fixed in formalin and exposed to the picric acid–mercuric chloride secondary fixation procedure, the time in the plasma stain may be reduced to 5 minutes.

Differentiation in phosphomolybdic acid is critical: this displaces the plasma stain from the collagen fibres. Differentiation of the collagen stain is needed to prevent the blue or green stain swamping the plasma stain.

Masson's trichrome is an excellent stain and is well worth practising. Timings vary, as usual, with tissue and fixatives. The method is slowed by the need for differentiation. Once this has been done on a few sections, the timings can be set and the process accelerated. An effective method that does not require differentiation is given next.

Lillie's hydrochloric acid–Biebrich scarlet–methyl blue method

Recipe

Solutions

Biebrich scarlet (CI: 26905)	1% in 1% acetic acid
Acetic acid wash	1% acetic acid
Methyl blue	0.5% in 0.5% HCl

Method

De-wax sections and take to water. Lillie recommended fixing tissue in formalin; if this is done, there will, of course, be no mercury pigment to remove. Stain nuclei with iron haematoxylin: 7.3.8.1.

Biebrich scarlet	5 minutes
Acetic acid wash	2 minutes
Methyl blue	5 minutes

Dehydrate in 95% alcohol, then absolute alcohol, dip in xylene and mount in DPX.

Results

Collagen, reticulin, fibrin, basement membranes	Blue
Muscle	Red, with cross striations a darker hue
Red blood cells	Scarlet
Cytoplasm	Grey–pink.

This is a rapid and very effective stain.

7.4.2 Fibrin Stains

Staining fibrin is easy; differentiating fibrin from newly formed collagen is less easy. It appears that the staining characteristics of fibrin change with the age of the fibrin: as it ages it takes on the staining characteristics of collagen. Lendrum's Martius–scarlet–blue (MSB) stain is an outstanding fibrin stain and allows clear differentiation between fibrin and mature collagen. Being certain that no fibrin has stained as collagen is probably impossible. Control sections are thus very necessary: staining serial sections with Lendrum's MSB and Masson's trichrome is recommended.

7.4.2.1 Lendrum's MSB Stain.   MSB stands for <u>M</u>artius yellow-brilliant crystal <u>s</u>carlet–soluble <u>b</u>lue, or <u>M</u>artius–<u>s</u>carlet–<u>b</u>lue. Lendrum recommended fixation of tissue in formol-sublimate (formalin plus mercuric chloride) and suggested that prolonged fixation aided the stain without making the tissue more difficult to section. Lendrum also recommended that, if formalin alone has been used as a fixative, the sections should be soaked in trichloroethylene for 48 h, followed by 24 h in 3% mercuric chloride in picric acid. The sections should then be washed and the mercury pigment removed as usual. Nuclear staining may be done with either Weigert's iron haematoxylin or the Celestine blue-haemalum stain. We shall assume that Weigert's haematoxylin will be used.

Recipe

Solutions

Martius yellow (CI: 10315)	0.5% in 95% alcohol containing 2% phosphotungstic acid
Brilliant crystal scarlet (CI: 16250. This dye is also known as Acid Red 44, crystal scarlet and Ponceau 6R. Lillie gave the CI as 16290 but this seems to refer to a different dye: Acid Red 41)	1% in 2.5% acetic acid
Soluble blue (CI: 42755, also known as aniline blue)	0.5% in 1% acetic acid
Phosphotungstic acid	1% in distilled water
Acetic acid	1% in distilled water

Method

De-wax and take sections to water.
Stain nuclei with Weigert's iron haematoxylin: see section 7.3.8.1.
Differentiate the nuclear stain in 0.5% HCl in 70% alcohol.
Wash well in tap water for 5 minutes.
Rinse in 95% alcohol.
Martius yellow stain for 2 minutes.
Rinse in distilled water.
Brilliant crystal scarlet 6R stain for 10 minutes.
Rinse in distilled water.
Differentiate in 1% phosphotungstic acid: continue until the collagen is unstained.
Rinse in distilled water.
Soluble blue stain for 10 minutes.
Rinse in 1% acetic acid.
Blot with filter paper.
Absolute alcohol   1 minute; 2×
Dip in xylene and mount in DPX.

Results

Nuclei	Black
Red blood cells	Very bright, clear yellow
Fibrin	Red
Collagen	Blue

The method stains fibrin very well but not entirely specifically. The colour contrast is excellent. Once again: do not over-stain.

7.4.3   Reticulin Fibres

Reticulin fibres contain type III collagen but are not well demonstrated by conventional stains for collagen. The fine collagen fibres are associated with sugar moieties in the form of proteoglycans. These reducing sugars allow impregnation with silver. This is sometimes referred to as "silver staining" but purists point out that the process is not really one of staining but one of impregnation and is shown in Figure 7.5. If the word "stain" is used in this section, with respect to silver, it will be taken to imply impregnation with silver.

Lillie divided reticulin fibres into three groups:

1. Pre-fibrous collagen reticulin. Newly formed collagen stains as reticulin.
2. Stromal reticulin. Organs such as lymph node, spleen and bone marrow have an extensive reticulin fibre skeleton. Reticulin fibres are found in the inter-alveolar septae of the lung.
3. Basement membrane reticulin. Basement membrane contains type IV collagen, as well as reticulin fibres, which are concentrated in the lamina reticularis. Basement membrane stains well with silver methods.

Points to note before beginning to use methods of silver impregnation

These methods involve the preparation and use of ammoniacal silver salts that contain the complex ion $Ag(NH_3)_2^+$. Such salts can be oxidised to silver nitride, Ag_3N, which is explosive. Silver nitride occurs as black crystals. Conversion of the ammoniacal silver to nitride can occur in deposits left on glassware, which acquires a brownish appearance, especially if the glassware is exposed to sunlight. Safety precautions must be observed.

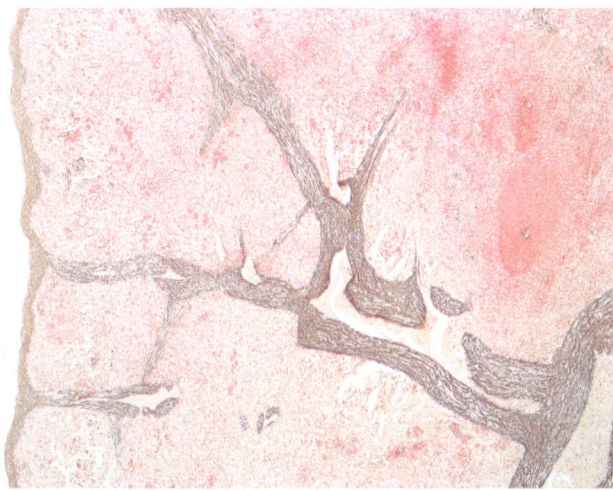

Figure 7.5   Reticulin stain. Reticulin stain (dark brown/black) in the spleen showing the underlying architecture that is not apparent with a routine stain.

1. All ammoniacal silver solutions should be prepared immediately before use.
2. Scrupulously clean glassware must be used at all times.
3. Ammoniacal silver solutions must not be exposed to sunlight.
4. Ammoniacal silver solution must be disposed of by adding an excess of dilute HCl or sodium chloride to precipitate silver chloride before discarding.

All this seems very worrying! However, if reasonable precautions are taken, the silver impregnation methods may be used entirely safely. It should be noted that these methods have been in regular use since about 1900.

Preliminaries

Formalin-fixed material should be used.

Sections should be stuck firmly to the slides using the albumen adhesive. The high pH of the stain tends to loosen the sections from the slide unless they are stuck firmly.

All solutions, except the gold chloride (if this is used: see below), should be made up fresh (Jacoby F, personal communication to RLM, 1976).

Recipe

Solutions

Acidified permanganate	47.5 mL 0.5% potassium permanganate, plus 2.5 mL 3% sulphuric acid
Oxalic acid	1% in distilled water
Ferric alum	2.5% in distilled water
Formalin	10% neutral formalin. Measure the pH. Use buffered formalin: see Chapter 5
Sodium thiosulphate	5% in distilled water
Gold chloride	0.2% in distilled water
Neutral red	1% in distilled water

Silver solution

This requires careful preparation. Ensure that all the glassware is perfectly clean and rinsed out with distilled water.

Solutions required:

Silver nitrate	10.2% in distilled water
Ammonia solution	SG: 0.880, "880 ammonia"
Sodium hydroxide	3.1% in distilled water

Take 5 mL silver nitrate solution and place in a small conical flask. Add the ammonia from a burette, drop by drop, noting that a white precipitate forms and then begins to disappear. Continue adding the ammonia, rotating the flask to mix the contents, until the precipitate has JUST disappeared. The solution should not be cloudy but should appear just off clear. Add 5 mL sodium hydroxide: the precipitate reappears. Add a little more ammonia until, again, the precipitate JUST disappears. Place the solution in a clean measuring cylinder and make up to 50 mL. Getting this process right is the key to the successful use of the method.

Method

De-wax sections and take to water.

Oxidise for 1–5 minutes in acidified permanganate.

Rinse in distilled water.

Bleach in oxalic acid until colourless: this take only a few moments.

Rinse in distilled water.

Wash in running tap water for 10 minutes.

Sensitise in ferric alum solution for 15 minutes.

Wash in distilled water 3×.

Stain in silver solution for 10–30 minutes: sections will become transparent.

Wash in distilled water $3\times$.

Reduce in formaldehyde solution for 1–2 minutes.

Wash in tap water for 2 minutes.

Rinse in distilled water.

Tone in gold chloride solution for 1–2 minutes: sections will take on a purple colour.

Rinse in distilled water.

Fix in thiosulphate for 5 minutes.

Wash well in tap water for 5 minutes.

Stain in neutral red for 1 minute.

Dehydrate, dip in xylene and mount in DPX.

Results

Reticulin fibres, basement membrane, some mucins	Black
Cytoplasm and collagen	Purple–brown
Nuclei	Pink–red

Points to note

This seems a complicated method! So it is, but if carried out carefully, good staining of reticulin is produced. Like all silver methods, it can fail on occasion, but this is a reliable method. The gold chloride toning is optional: the step may be omitted if you wish. Without the gold toning the collagen and cytoplasm in general appears a pale yellow–brown. Adding collagen stains and stains for elastic fibres to the method is possible but complicates it even further. Staining serial sections for collagen and reticulin is recommended. Sections of spleen provide an excellent positive control for staining reticulin fibres: such sections should be "put through" with the test material.

7.4.4 Staining for Elastic Fibres

Several reliable methods are available. Orcein, originally extracted from moss but now synthesised and available in pure form, is an excellent stain. For other methods, see the manuals listed in the Bibliography.

7.4.4.1 *Orcein Staining of Elastic Fibres*

Recipe

Solution

Orcein: 1 g dissolved in 10 mL 80% alcohol. When dissolved, add 1 mL conc. HCl.

Method

De-wax and take sections to 70% alcohol (NOT to water).

Stain in orcein for 30 minutes in a closed jar. The stain evaporates and this method is not suitable for use on glass rods.

Wash in 70% alcohol.

Examine with the staining microscope. Some collagen fibres will be stained in addition to the elastic fibres.

De-stain the collagen in 1% HCl in 70% alcohol.

Wash in 70% alcohol.

Dehydrate, dip in xylene and mount in DPX.

Results

Elastic fibres	Deep brown
Nuclei	Light brown
Collagen	Light brown if the differentiation has not been adequate

Points to note

Counter-staining for collagen or adding a background plasma stain is often recommended. This is not necessary: the architecture of the tissue can be made out clearly in the orcein-stained sections.

7.5   STAINING FOR CARBOHYDRATES AND MUCOSUBSTANCES

7.5.1   The Complex Chemistry of Mucosubstances

Histologists speak of mucin, histochemists refer to mucosubstances and mucopolysaccharides, biochemists prefer glycosaminoglycans and other long words. Here, we shall use as simple a nomenclature as possible. For our purposes, we may think of three groups of compounds that form mucosubstances.

1. Mucopolysaccharides: complexes of polysaccharides and protein, with the former outweighing the latter.
2. Mucoproteins: complexes of polysaccharides and protein, with the latter outweighing the former.
3. Mucolipids: complexes of polysaccharides and lipids.

We shall ignore the mucolipids and focus on the other two groups. Mucosubstances are found as mucins: the epithelial mucins produced by, for example, the gastric epithelium, goblet cells of the colon and respiratory tract and by some of the salivary glands; and connective tissue mucins, which form the ground substance of connective tissue.

A little more detail is now needed.

Connective tissue mucins **comprise:**
Acid mucopolysaccharides. These comprise:

- The carboxylated acid mucopolysaccharides: for example, hyaluronic acid found in the umbilical cord and in the skin. In these, the monosaccharide units carry COOH (carboxyl) groups.
- The sulphated and carboxylated acid mucopolysaccharides: for example, chondroitin sulphate found in cartilage, in the wall of the aorta and in heart valves, and heparin found in the granules of mast cells.
- The carboxylated and sulphated acid mucopolysaccharides: for example, keratosulphate found in cartilage and in intervertebral discs.

Epithelial mucins **(which might be known as** *epithelial mucosubstances*) **comprise:**

- Neutral mucoproteins: for example, the mucin produced by the stomach and prostate.
- Carboxylated mucoproteins: the sialomucins; these carry carboxyl groups and are produced by the small intestine.
- Carboxylated and sulphated mucoproteins: the sialated sulphomucins produced by the colon.

In addition, *glycogen*, a polysaccharide with no protein attached to the molecule and which is the only storage polysaccharide found in mammalian tissue, should be included to complete our list of carbohydrates and mucosubstances that are routinely stained in histological sections. Details of the chemistry of these compounds can be found in textbooks of biochemistry. Sialic acid is a carboxylated nine-carbon monosaccharide and forms part of the structure of the sialomucins and sialated sulphomucins. Sialic acid residues are also important components of membrane glycolipids and act as receptor sites.

The stress placed on the presence of acid radicals is explained by their importance in controlling the staining reactions of these complicated compounds. Both the carboxyl and the sulphate radicals can release a proton (they are acid radicals), and thus acquire a negative charge. The release of the proton depends on the pK_a of the group and the pH of the medium. By adjusting the pH of the staining solution these acid groups can be stained for differentially. This allows different sorts of mucins to be identified in sections of tissue.

The material secreted by glands that produce mucins is called mucus. Note that mucus is a noun and that the adjective is mucous. Thus, mucous glands or mucous cells produce mucus. A mucous membrane is one that secretes mucus. There is more to mucus than mucin! Other components, including materials secreted by serous cells, may mix with the secretions of mucous cells to produce the mucus that forms on the surface of the epithelium. Mucus that we cough up after a severe cold contains mucins to be sure, but also debris of dying and dead leucocytes and bacteria. It may be yellow or green. The green colour is said to be due to the copper-containing myeloperoxidase enzyme found in neutrophils.

Drury and Wallington (1980) listed methods for staining carbohydrates (glycogen) and mucosubstances. Their list forms the basis of the following, short, discussion.

1. The $1:2$ glycol linkage of monosaccharides can be broken and the alcohol groups converted to aldehyde groups by oxidising agents—the compound used in periodic acid. The aldehyde groups can react with decolourised basic fuchsin (Schiff's reagent) and restore the colour to the reagent. This is a true histochemical technique: it is specific for aldehyde groups. The method is discussed below.
2. The acid groups (carboxyl and sulphate) can be stained for with basic dyes. Alcian blue is such a dye: the method is discussed below.
3. The acid groups can be demonstrated by their effect on metachromatic dyes, such as thionin blue. Metachromasia is discussed in Chapter 8. We shall not discuss it further here.
4. A number of older methods, including staining with iodine and carmine, can be used.
5. Silver impregnation methods can be used: here, the aldehyde groups that are produced by oxidation of the alcohol groups, see above, act as reducing agents. See Chapter 8 on silver impregnation techniques. See also section 7.4.3 on staining for reticulin fibres: what are being stained are the polysaccharides associated with the type III collagen fibres.

It is interesting to note that glucose is classically described as a "reducing sugar", and so it is under certain circumstances. In the old days of clinical chemistry, glucose was detected in urine by boiling urine with freshly made Fehling's solution. The sugar reduced the soluble cupric salt to reddish-brown cuprous oxide. The reduction is carried out by aldehyde groups. The aldehyde groups were produced, so to speak,

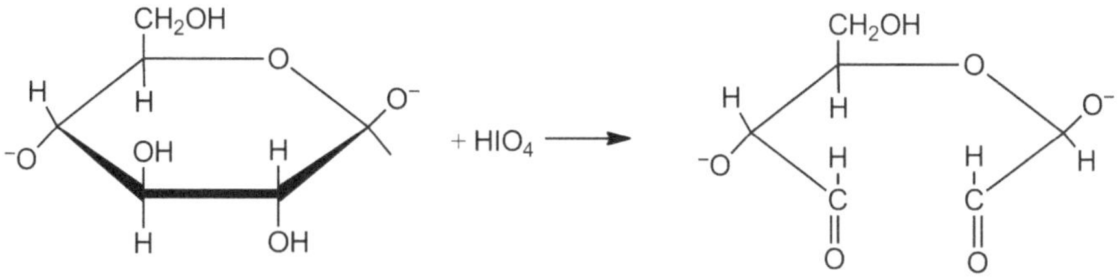

Scheme 7.3 Reaction between periodic acid and the dehydroglucose residue of a polysaccharide.

by the strongly alkaline medium used: the monosaccharide was converted to an enol with reactive aldehyde groups. Given that glucose is a "reducing sugar", why is periodic acid needed as an oxidant to produce aldehyde groups before Schiff's reagent can be re-coloured? This suggests that the aldehyde groups are not immediately available. They are not. Glucose is described as an aldohexose: six carbon atoms and an aldehyde group. If the formula for glucose is written as a straight chain of six carbon atoms then C1 carries an aldehyde group. The failure to re-colour Schiff's reagent suggested that the aldehyde group was not available and this led to the suggestion that, as we all know, glucose exists as a ring of five carbon atoms with C6 attached to C5. What has happened is that C1 has become linked to C5 to form what is described as a hemiacetyl linkage. The reaction between periodic acid and the dehydroglucose residue of a polysaccharide is shown in Scheme 7.3. See Pearse for further details. The message is: aldoses are reducing sugars in strongly alkaline solutions because of enol formation. Monosaccharides, like glucose, occur as rings and the aldehyde group is not available for reaction with Schiff's reagent. Aldehyde groups can be made available by oxidation with periodic acid.

7.5.2 Why should One Wish to Stain Carbohydrates and Mucosubstances?

Having struggled with the biochemistry, it might be useful to consider why glycogen and the diverse mucosubstances are looked for in histological sections.

7.5.2.1 Glycogen. As noted above, glycogen is the only form in which carbohydrate can be stored by mammalian cells. Storage is particularly important in the liver and muscle. Glycogen can be seen as a reserve, which can be drawn upon if the supply of glucose fails. Substances that interfere with oxidative phosphorylation also lead to depletion of glycogen, which is broken down to glucose and used in less efficient pathways of adenosine triphosphate (ATP) production. Liver cells are depleted of glycogen in a number of diseases of the liver, including hepatitis; myocardial cells are depleted of glycogen if the blood supply to the myocardium is impaired. Depletion of glycogen might be seen as a sign that a cell is "under pressure". Glycogen appears in cells in increased amounts in a number of genetically determined diseases. Not unreasonably, these are known as glycogen storage diseases. A number of different diseases of this type have been described but all include a lack of enzymes involved in the breakdown of glycogen. Glycogen appears in increased amounts in the cells of the distal convoluted tubule of the kidney in cases of diabetes mellitus. It also appears in increased amounts in the beta cells, the insulin-producing cells of the pancreas, in cases of diabetes mellitus. The reasons underlying these abnormal increases of glycogen in certain cells seem to be incompletely understood.

7.5.2.2   *Mucosubstances*

Epithelial mucins

Inflammation of mucosal surfaces (that produce mucus from the surface columnar cells, from goblet cells of the surface epithelium or from submucosal glands) leads to an increase in production of mucus. Thus, a common cold leads to an increase in mucus production by the epithelial cells of the nasal cavity; bronchitis leads to an increase in mucus production by goblet cells of the airway epithelium and by submucosal glands. Ulcerative colitis is associated with an outpouring of mucus from the colonic epithelium. Chronic irritation and inflammation, as seen in chronic bronchitis of smokers, leads to hypertrophy of the submucosal glands of the airways. Changes in the chemical composition of the mucins produced also occur: whether these changes represent a defensive response is unclear.

Epithelial mucins are produced in large amounts by certain tumours. Certain tumours of the stomach and colon can comprise mainly a mass of mucoid material. Tumours of the breast can also, albeit less often, produce large quantities of mucin. The cystadenoma tumour of the ovary produces very large amounts of mucin.

Production of epithelial mucins is, unsurprisingly, reduced when metaplasia of a columnar epithelium leads to its replacement with a squamous epithelium.

Connective tissue mucins (mucosubstances)

A large number of genetically determined diseases are characterised by an increase in the amount of mucosubstances found in connective tissue. These diseases are "storage diseases" and form a parallel with the glycogen storage diseases. The many diseases that fall into this category can be divided into:

- Mucopolysaccharides storage diseases;
- Glycolipid storage diseases;
- Phospholipid storage diseases (not a mucosubstances disease);
- Cholesterol ester and triglyceride storage diseases (not muco-substances diseases);
- Glycoprotein storage diseases.

Details of these conditions will be found in textbooks of medicine or pathology: see Bibliography.

Mucosubstances appear in abnormal amounts in areas of mucoid degeneration of tumours: for example, a myxoma is a connective tissue tumour, which presumably began with mutation of fibroblasts and which is characterised by its mucin production. The appearance of mucin in what had been regarded as a benign fibroma is a sign that malignant change is taking place. A good example of a benign swelling that is filled with mucin is the ganglion, which occurs, commonly, on the dorsal aspect of the wrist. This is a benign tumour of the synovium of the tendon sheath of one of the extensor muscle of the fingers. Treatment used to be said to comprise striking the ganglion with a heavy book (family Bible, textbook of anatomy), which would burst the swelling and disperse the mucinous fluid. Surgery offers a more reliable, if less dramatic, cure.

Myxoedema, the swelling of connective tissue seen in cases of thyroid deficiency amongst adults, also involves the deposition of abnormal amounts of mucosubstances. Rather oddly, localised swelling known as pre-tibial myxoedema occurs in hyperthyroidism. Why this should occur seems to be unknown. Interestingly, this localised form of myxoedema involves the deposition of hyaluronic acid, the same mucosubstance as is deposited in the generalised myxoedema of hypothyroidism.

Lastly, mucoid degeneration can occur in bone marrow, adipose tissue and cartilage in cases of malnutrition.

It will be clear that there are many reasons why a histopathologist might wish to look at mucosubstances. The following account focuses only on glycogen and the epithelial mucins. Only a few of the many methods that are available are described. These methods will allow successful staining of mucosubstances. More detailed accounts may be found in textbooks of histochemistry: see Bibliography.

7.5.3 Staining for Glycogen

Glycogen may be stained by means of the periodic acid Schiff (PAS) method. This method has already been mentioned: it was noted that it is a true histochemical method in that it is specific for aldehyde groups. These groups are produced by oxidation. Glycogen and all the epithelial mucins may be stained by the PAS method. Looking ahead for a moment, it is interesting to note that none of the acid mucopolysaccharides (the connective tissue mucins) are stained by this method.

As with all histochemical methods, control sections are essential. A negative control for PAS staining for glycogen can be produced by exposing the section to diastase, found in saliva and is the explanation for the common observation that, as bread is chewed, it becomes sweeter. In fact, there is nothing wrong with using saliva to produce the negative control, though a solution of 1% diastase in distilled water is more convenient if many sections are being handled. A very efficient negative control can be produced by staining adjacent serial sections: one with PAS *without* the diastase control and one with PAS *with* the diastase control.

Glycogen used to be thought to disappear from sections fixed with formalin and fixation with alcohol used to be recommended. This is unnecessary: as long as small pieces of tissue are taken for fixation, formalin is an adequate fixative. But better fixation is provided by a mixture of formalin and alcohol:

Formalin (40%)	10 mL
Absolute alcohol	80 mL
Distilled water	10 mL

Glycogen is a polysaccharide and is not actually fixed by any fixative: recall that fixatives fix protein (see Chapter 5). But fixing protein leads to the glycogen being, as it were, held in its place in sections. All would be well except for the fact that, as the fixative diffuses into the tissue, the glycogen seems to stream along with it. This produces an artefact: the glycogen piles up at one side of the cell. This artefact can be prevented by fixing with a variant of Bouin's fixative (alcoholic Bouin's) at 4 °C. This fixative is made up as follows:

Saturated alcoholic solution of picric acid	75 mL
Formalin (40%)	25 mL
Glacial acetic acid	5 mL

It will be recalled that Bouin's fixative is a curiously poor fixative of kidney tissue. For general use, the alcoholic formalin described above should be preferred.

7.5.3.1   PAS Method

Recipe

Solutions
Schiff's reagent

Add 1 g of basic fuchsin to 200 mL boiling distilled water, allow to dissolve and cool. Filter. Bubble sulphur dioxide through the solution until the pink–red colour disappears. Sulphur dioxide is conveniently supplied in small cylinders; it is an irritant and should be used in a fume cupboard. Exposure to sulphur dioxide causes eye irritation and low concentrations cause bronchospasm in those suffering from asthma. The colourless solution should be kept in a refrigerator. If the decolourisation is incomplete (a pink tinge remaining) then add 1 g activated charcoal, shake and filter. The solution is said to keep fairly well but making up a new batch when it is needed is a good idea.

Periodic acid 1% in distilled water.

Method
De-wax and bring sections to water.

Periodic acid	5 minutes
Wash in running tap water	5 minutes
Rinse in distilled water.	
Schiff's reagent	15 minutes

Wash in running tap water for 10 minutes.

If a nuclear stain is required: stain with Weigert's iron haematoxylin: see section 7.3.8.1. The nuclear stain is not absolutely necessary: the architecture of the tissues can be made out sufficiently clearly without it.

Dehydrate, dip in xylene and mount in DPX.

Results
Many components of normal tissues stain with the PAS reaction. These include glycogen, basement membrane, cartilage ground substance, epithelial mucins, colloid of thyroid follicles and zymogen granules of cells of pancreatic acini. But acid mucopolysaccharides stain faintly, if at all. The distribution of staining shows that mucosubstances in many tissues are rather mixed.

Negative controls
1. Repeat, omitting the periodic acid bath: no staining should occur.
2. The diastase control for glycogen.

Solution
1% commercial diastase (derived from malt) in distilled water.

Method
De-wax and bring sections to water.

Diastase solution at 37 °C (in an incubator: put the diastase solution in the incubator to warm up to 37 °C before immersing the sections) for 30 minutes.

Then perform the PAS staining as described above.

Results
Of all the components listed above as being stained by the PAS method, ONLY glycogen is NOT stained after use of diastase. When staining serial sections, a diastase control, PAS without diastase, diastase control and a control by omitting periodic acid is advised.

7.5.4 Alcian Blue Methods

Alcian blue is a most important stain for mucosubstances. Varying the pH of the staining solution allows (1) acid groups to be demonstrated and (2) carboxyl and sulphate groups to be differentiated, as shown in Figure 7.6. Staining with Alcian blue at pH 0.5 can be combined with staining with Alcian yellow at pH 2.5: this allows a very

Figure 7.6   Alcian blue at different pH. On the left, the stain was done at a pH of 1.0 and counter-stained with PAS; a quite different appearance is seen if the stain is done at a pH of 2.5 (right). Pictures courtesy of Christine Ruehl-Fehlert.

attractive differentiation of acid groups to be produced. Alcian blue can also be combined with PAS staining: this allows differentiation of neutral and acid mucosubstances. Alcian blue is particularly effective for staining of epithelial mucins.

7.5.4.1 Alcian Blue–Alcian Yellow Staining

Recipe

Solutions

Alcian blue 8GX (CI: 74240)	1 g dissolved in 100 mL 0.5 M HCl
Alcian yellow (CI: 12840)	1 g dissolved in 100 mL 3% acetic acid
0.5 M HCl	
Neutral red	0.5% in distilled water

Method

De-wax and bring sections to water.
Alcian blue 5 minutes
Rinse in 0.5 M HCl.
Wash in tap water.
Alcian yellow 5 minutes
Wash in tap water.
Neutral red 1 minute
Wash in tap water.
Dehydrate, dip in xylene and mount in DPX.

Results

Epithelial sialomucins (COOH groups, staining at pH 2.5)	Yellow
Epithelial sulphated mucins (COOH and SO_4 groups, the latter staining at pH 0.5)	Pale blue
Mixtures	Green–blue
Nuclei	Pale pink–red

7.5.4.2 Further Uses of Alcian Blue. Alcian blue may be used at pH 2.5: 0.5 g Alcian blue dissolved in 100 mL 3% acetic acid; or at pH 1.0: 1 g Alcian blue dissolved in 100 mL 0.1 M HCl. The methods are similar to that described for Alcian blue–Alcian yellow staining. Drury and Wallington recommended 10–30 minutes in the stains when using Alcian blue alone at these pH levels. This is longer than that recommended when using the Alcian blue–Alcian yellow method. Examining the sections with the staining microscope will allow the optimal timing to be set. If it is desired to combine Alcian blue at pH 2.5 with PAS staining, sections should be stained for 10–30 minutes in the Alcian blue stain, washed with distilled water and then stained with PAS as described in section 7.5.3.1. Alcian blue staining is easy to do and produces reliable results.

7.5.5 Further Remarks on the Staining of Mucosubstances

A number of other methods for staining mucosubstances are available. Further methods for differentiating different types of mucosubstances and for producing negative controls are also available. Information on the following methods will be found in books on histochemistry listed in the Bibliography.

- Use of metachromatic dyes;
- Use of varying concentrations of electrolytes to allow differentiation of mucosubstances;
- Use of methylation and saponification to allow differentiation of mucosubstances.

7.6 STAINING FOR LIPIDS

The nomenclature of fats or lipids is almost as distressing as that of carbohydrates and mucosubstances. The following classification is based on that provided by Drury and Wallington.

Simple lipids
1. Neutral fats: triglycerides, the main storage form of fat found in adipose tissue.
2. Ester waxes: cholesterol esters found in cells of the adrenal cortex.

All simple lipids are insoluble in water but soluble in organic, non-polar solvents, and thus disappear from tissue processed for paraffin wax sections. Simple lipids are preserved in frozen sections and can be coloured by the use of lysochromes, which dissolve in the fat. See Chapter 9.

Compound lipids
1. Phospholipids. Recall that in triglycerides all the OH groups of glycerol, a polyhydric alcohol, are esterified by fatty acid. In the phospholipids only two of the OH groups of glycerol are esterified in this way, the third is linked *via* phosphoric acid to a

nitrogenous base. In the case of lecithin this nitrogenous base is choline.

2. Sphingolipids. These compounds are all found in the nervous system and can be subdivided into sphingomyelins, cerebrosides, gangliosides and sulphatides.

Derived lipids
1. Fatty acids.
2. Sterols. These are produced by hydrolysis of ester waxes. Cholesterol is the obvious example but the steroid hormones, including testosterone, progesterone and the adrenocortical hormones, should be recalled. Bile salts also fall into this category.

7.6.1 Why should We Wish to "Stain" Lipids?

The reason for the quotation marks around "stain" is that lipids are not stained in the sense that chromatin is stained in an H&E section, but coloured by lysochromes that dissolve in the lipid but are not chemically bound to it.

From the toxicologist's point of view, perhaps the most important reason for staining (colouring!) lipids is that fatty degeneration of the liver is produced by a wide range of chemicals. Chloroform and carbon tetrachloride are the classical examples. Fatty degeneration is characterised by the appearance of globules of lipid in the hepatocytes. Fatty degeneration is also produced by anoxia. This is seen in the myocardium if its blood supply is impaired. This also occurred in the "old days" before vitamin B12 was introduced for the treatment of pernicious anaemia. The lack of oxygen (due to the lack of red blood cells) led to fatty degeneration of the left ventricle and its surface became spotted with fat. This led pathologists to describe the change as "thrush breast heart": pathologists seem to be less ornithologically orientated today.

Atheroma of the tunica intima and tunica media of arteries involves the accumulation of fat. Breakdown of the fat leads to the formation of fatty acids and cholesterol. The former provoke an inflammatory reaction. Cholesterol , like other lipids, is dissolved during processing for paraffin sections but "where it was" can often be recognised by clefts in the tissue: cholesterol clefts.

Fat can also accumulate in the connective tissue stroma of organs and infiltrates along connective tissue planes. The heart may be affected in this way. The pancreas can also accumulate a lot of fat in its connective tissue stroma, sometimes to the point where glandular tissue is difficult to find. A number of rather rare lipodystrophies exist: in these conditions abnormal deposition of fat occurs. These are of more interest to the clinical pathologist than to the experimentalist.

7.6.2 Histological Appearance of Lipids

Lipids are often recognised in paraffin wax sections by the spaces they leave behind when dissolved during processing. Thus, the cells of adipose tissue appear as great empty balloons with just a trace of cytoplasm and a flattened nucleus. Fatty degeneration of hepatocytes leads to the cells becoming swollen with fat: in paraffin wax sections they appear to be full of holes. Some cells show many small holes due to smaller globules of fat being dissolved during processing. This is a very typical picture of the fatty degeneration seen in the livers of alcoholics. The appearance of the cells looks almost as though areas of the liver were being converted into adipose tissue. One technique, which does allow fat to be preserved and seen in paraffin sections, is provided by the use of Flemming's chromium trioxide–osmium tetroxide fixative. Lipids are blackened by the osmium tetroxide and this is not lost as the tissue is carried through the processing to paraffin sections. Flemming's fixative is discussed in Chapter 5. It is a classical cytological fixative but is tricky to use and not very useful for ordinary histological work. Osmium tetroxide is, of course, widely used as a fixative/stain in electron microscopy. Frozen sections provide the best basis for demonstrating lipids.

7.6.3 Fixation of Tissue Prior to Preparation of Frozen Sections

No fixative-containing organic solvents should be used. The best fixative is probably Lillie's formalin–calcium acetate mixture:

Formalin (40%)	100 mL
Distilled water	900 mL
Calcium acetate (monohydrate)	20 g

This mixture does not require buffering and the calcium aids in the preservation of the lipid.

The essence of lysochrome colouring is that the lysochromes partition to the fat away from their solvents. We have realised that such solvents should be avoided: they dissolve the fat. 70% alcohol is a suitable compromise and has been widely used as a solvent for lysochromes, but a better solvent is provided by triethyl phosphate.

7.6.4 Oil Red O–triethyl Phosphate Method

Recipe

Solution

   Oil red O (CI: 26125): 1 g added to 100 mL 60% triethyl phosphate (triethyl phosphate made up in distilled water). Place in the paraffin wax in an oven at 56 °C for 3 h. Stir now and again. Filter whilst hot, cool and filter again.

Method

   Rinse sections in 60% triethyl phosphate.
   Colour in the Oil red O solution for 15 minutes.
   Wash in triethyl phosphate.
   Rinse in distilled water.
   (The fat is now coloured bright red. A nuclear counter-stain may be applied. Light staining with alum haematoxylin is recommended. Be careful about differentiating the haematoxylin stain in acid alcohol: this will remove the fat stain. Ehrlich's haematoxylin, which works well as a progressive stain, may be used: the bluing in tap water is, of course, necessary but the differentiation can be avoided. Stain the nuclei lightly.)
   Mount in an aqueous mountant. You cannot use DPX because, of course, that mountant requires prior clearing in organic solvents. See Chapter 6.

Results

   Fat   Bright red
   Nuclei   Pale blue–purple

7.6.5 Other Methods

Special methods for staining phospholipids and deposits of cholesterol are available. Extraction of lipids allows negative controls to be generated. Details of these methods may be found in the latest edition of Carleton: *Carleton's Histological Technique* by Drury and Wallington; see Bibliography.

7.7 CELLOIDIN AND FROZEN SECTIONS

7.7.1 Celloidin Sections

Celloidin sections may be stained after being attached to slides or as loose sections. Celloidin sections that have been attached to slides may be stained with the celloidin in place or after the celloidin has been removed by treatment with a 50/50 mixture of absolute alcohol and ether. Staining of loose sections is the usual method and this is given here. The celloidin is not removed; to ensure this, the use of absolute alcohol is avoided.

Celloidin sections are, in comparison with paraffin wax sections and, especially, frozen sections, rather robust and can be moved easily from one dish of fluid to another using either a piece of bent glass rod (thin glass rod with an inch or so at one end bent to 45°) or a section lifter. The following method is from Carleton (1957): he found it produced flatter sections than other methods. It is one of many methods: see manuals of technique listed in the Bibliography.

Celloidin sections should be stored in 70% alcohol. Assuming that this is the case:

50% alcohol for 5 minutes.

Wash in water.

Stain in Ehrlich's haematoxylin as described in section 7.3.3.1 for paraffin wax sections.

Differentiate in 1% HCl in 70% alcohol.

Blue the sections in tap water or tap water substitute.

Stain in eosin for 5 minutes. Penetration is slower than with wax sections.

Wash in water.

95% alcohol for 3 minutes.

Place sections in a mixture of equal parts of xylene, chloroform and absolute alcohol for not longer than 5 minutes.

Transfer sections to a dish of pure beechwood creosote for not longer than 7 minutes.

Float the section onto a slide.

Blot with hard toilet paper (the sort that comes in boxes rather than as rolls, do not use soft toilet paper. The older books recommend the use of cigarette paper: more available in laboratories then than now).

Pour xylene over the slide, drain and repeat. Do this quickly as the xylene may cause the section to crease.

Blot rapidly with the toilet paper.

Mount in DPX. Use plenty of DPX: it will be absorbed by the celloidin and tend to retract under the coverslip.

Carleton noted that diluting the stains with distilled water and extending the staining period improved the results.

7.7.2 Frozen Sections

It is probably easiest to attach frozen sections to slides before staining. Once attached, the sections can be treated in the same way as paraffin wax sections with, of course, the exception that the sequence of xylene and the descending grades of alcohol will not be needed to take the sections to water. Once stained, the sections should be dehydrated and cleared with xylene and mounted in DPX in the usual way. Staining times can be reduced in comparison with paraffin wax sections.

If frozen sections are to be stained loose then great care needs to be taken in transferring the sections from one dish of fluid to another. Davenport recommended the use of a perforated section lifter: one can be made from an ordinary section lifter by drilling some holes in its blade; remember to clean off any rough edges after drilling as these could catch on the section. The process of staining is as for paraffin wax sections that have been taken to water.

7.8 CONCLUSION

Choosing your staining method is a very important part of your study as the stains chosen can enhance the scope and specificity of your study. Further options for specific stains and forms of labelling are discussed in the following chapters.

CHAPTER 8

The Theoretical Basis of Histological Staining

This chapter sets out some, but not all, of the theoretical background to histological staining. We have assumed that the reader would wish to know something about the stains he or she uses and how they work. In particular the reader may be interested in differential staining: why some stains colour one component of tissues, whilst other stains colour other components. It has not been possible to go into the chemistry in great detail but a number of structural formulae have been included to give the reader some sense of the nature of the chemicals he or she will be using every day. The effects of varying the pH of staining solutions have been considered in some detail. This is important in everyday practice. The interesting phenomenon of metachromasia is discussed, as is the use of metallic impregnation.

8.1   INTRODUCTION

Much of histology depends on colouring components of tissues. Colouring is a good generic term that embraces staining, dyeing and other means of producing colours. I suppose that "staining" means much the same as "colouring" but staining has a more specialised meaning for the histologist. Why colouring of histological sections is

Histological Techniques: An Introduction for Beginners in Toxicology
By Robert Maynard, Noel Downes and Brenda Finney
© Maynard, Downes and Finney 2020
Published by the Royal Society of Chemistry, www.rsc.org

needed is easily explained by considering a glass rod in a beaker of water. The rod can be seen because it has refractive index that is different from that of the surrounding water. If the glass rod is placed in a bottle of disterene–plasticizer–xylene (DPX) mounting medium, it becomes invisible because the mounting medium has almost the same refractive index as the rod. If a rod of blue glass is placed in a bottle of mounting medium, it will be seen. Tissue that has been dehydrated, cleared and mounted in a medium with much the same refractive index as glass would be invisible to the microscopist. Histologists colour components of the tissue to allow them both to be seen and to be differentiated from one another.

The best account of colouring of tissues for histological examination has been provided by Baker in his *Principles of Biological Technique; a Study of Fixation and Dyeing* published in 1958 and reprinted in 1970: see Bibliography. Baker died in 1984: an account of his life and work, a little less laudatory than it might have been, has been provided by Willmer and Brunet (*Biographical Memoires of Fellows of the Royal Society*, 1985, **31**, 32–63). Baker devoted 157 pages to the subject of colouring of tissues; the account provided here is, in some ways, a summary of, or an introduction to, his work. Readers are encouraged to consult the original: not only is it well written but it contains information hardly available anywhere else.

The science of dyeing involves a great deal of organic chemistry. Baker dealt with this elegantly and at a level that can be followed by the biologist. Of course, a good deal more could be said by the expert organic chemist and some of the explanations provided by Baker might be seen by such an expert as superficial. For example, why, precisely, does a minor change in the molecular structure of a dye molecule alter the capacity of a solution of the dye to absorb light of certain wavelengths, and thus change colour? That it does is clear enough, but an explanation couched in terms of interactions between electromagnetic radiation of certain wavelengths and electrons of certain levels of activity might not mean much to one without the training (or perhaps the ability) needed to understand such advanced concepts. Those seeking such explanations will find them in advanced textbooks of organic chemistry. Finar provides a detailed account (see Bibliography).

A number of different processes can be used for colouring of histological sections of tissue. Table 8.1 is based on one provided by Baker.

Table 8.1   Dyeing and colouring.

Process	Example	Example of method
Dyeing	Staining of chromatin Staining of cytoplasm	Haematoxylin staining Eosin staining
Injection of suspended coloured particles into enclosed spaces	Colouring blood vessels	Injection of gelatin that has been coloured with carmine
Uptake of suspended coloured particles by cells	Colouring of macrophages	Injection of Indian ink
Dissolving a coloured substance in a component of tissue	Staining of lipids	Staining with Oil Red O
Absorption from solution of a coloured substance that is not a dye	Colouring of protein	Use of potassium tri-iodine
Local formation of a coloured substance that is not a dye	Identification of ferric iron	Colouring with potassium ferrocyanide
Local formation of a dye	Identification of aldehyde groups	Use of decolourised Schiff reagent

8.2   DYEING

Baker provided a useful definition of a dye, a definition not from the standpoint of the organic chemist, but from that of the histologist:

> *"A dye may be defined, for the purposes of microtechnique, as an aromatic, salt-like compound having these characteristics:*
>
> 1. *It ionises in the presence of water;*
> 2. *Either the cations or the anions are coloured (sometimes both);*
> 3. *The coloured ions are able to make chemical linkages with proteins and generally with other constituents of the fixed tissues of organisms (and in some cases with constituents of living cells as well);*
> 4. *When the coloured ions make their linkages with the tissues, they do not lose their colour, and generally they do not change it."*

The frequent use of the word "generally" suggests the occurrence of exceptions.

8.2.1   The Chemistry of Dyes

The definition given above will give us some clues as to the likely structure of dyes. We should, for example, expect that the aromatic compounds will carry groups that can be ionised. These groups will be capable of forming cations (positive charge) and/or anions (negative charge). Aromatic compounds contain carbon atoms arranged in rings. Benzene is the simplest example of an aromatic compound. It will be known that the position of the double bonds in the benzene ring cannot be specified with certainty and some authors do not draw in the bonds but represent the molecule by a hexagon containing a circle. The continuous redistribution of the bonding is described as resonance. Resonance involves the absorption of electromagnetic radiation. The wavelength of electromagnetic radiation that is most absorbed depends on the structure of the molecule: thus, different structures lead to different colours. The molecules of an aliphatic compound, such as ethanol, do not resonate and ethanol is colourless. Benzene is also colourless: the wavelengths that are absorbed are in the ultraviolet, to us invisible, range.

A chemical group that resonates and absorbs electromagnetic radiation in the visible range is described as a chromophore. A very good example is provided by the quinonoid ring. A simple structure containing this ring is parabenzoquinone (Scheme 8.1).

If we look at the dye pararosaniline (Scheme 8.2) many of the key points of dye chemistry can be illustrated.

Note that this is a three-ring molecule. Note that the quinonoid group is in place. The "para" does not have its usual chemical meaning here: it means that the dye was derived from rosaniline. The position of this group can vary as the molecule resonates (Scheme 8.3).

Note that the NH_2 group (the amine group) of the quinonoid group is charged. Nitrogen has a valency of three; if four bonds are shown attached to the nitrogen atom then that atom will be positively charged: it will have lost an electron. The NH_2 group is described as an auxochrome group. It is responsible for binding of the dye molecule to some negatively charged species in tissue and, importantly, it affects the colour of the dye. Anionic auxochromes include the sulphonic acid anion and the carboxyl anion (Scheme 8.4).

Scheme 8.1 Chemical structure of parabenzoquinone.

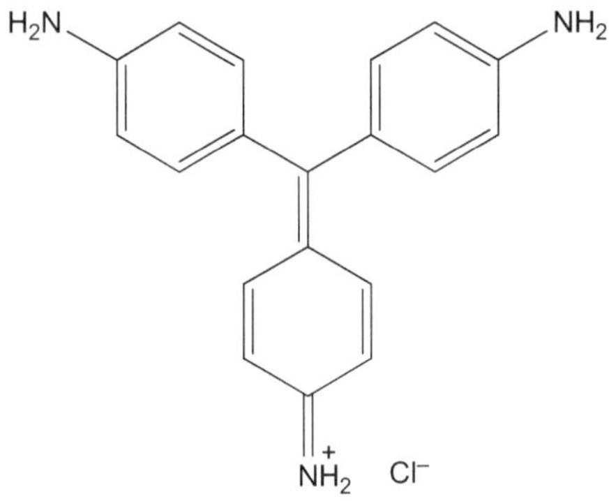

Scheme 8.2   Chemical structure of pararosaniline.

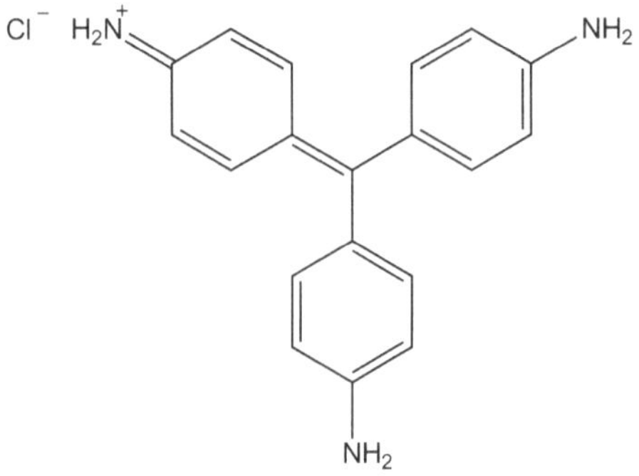

Scheme 8.3   Resonance form of pararosaniline.

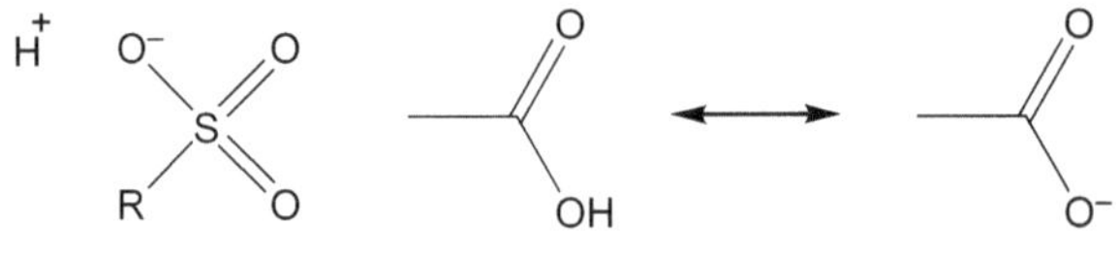

Scheme 8.4   Sulphonic acid (left) and carboxyl acid (right) auxochromes.

Some dyes carry cationic and anionic auxochromes. Basic fuchsine is actually a mixture of pararosaniline and rosaniline (Scheme 8.5). Acid fuchsine is sulphonated pararosaniline (Scheme 8.6).

8.2.2   Spectrophotometry Applied to Solutions of Dyes

Figure 8.1 (modified from Baker, *Principles of Biological Technique; a Study of Fixation and Dyeing*, 1958, p. 161) shows the transmission curve of basic fuchsine.

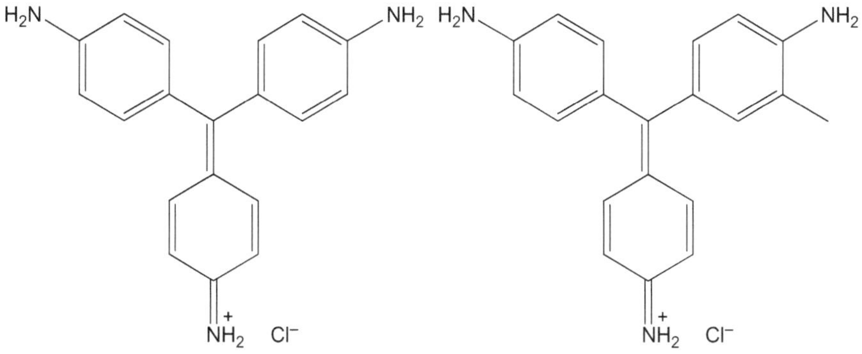

Scheme 8.5 Chemical structures of pararosaniline (left) and rosaniline (right).

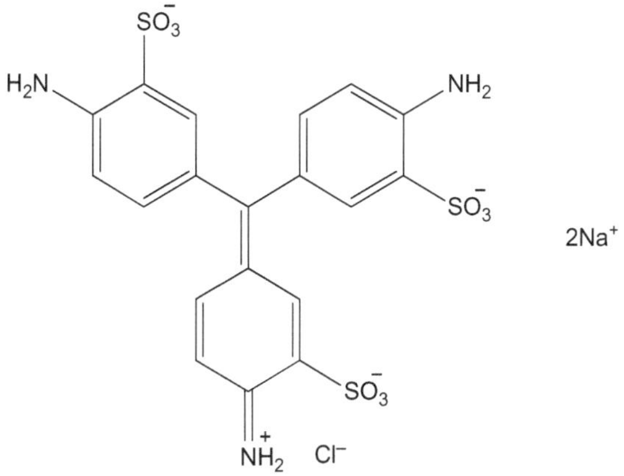

Scheme 8.6 Chemical structure of acid fuchsine.

This curve is useful to the histologist: it allows the colour likely to be imparted to tissue to be seen at a glance. The reciprocal of transmission (absorbance) or its logarithm may also be plotted. Gurr (1971) provided absorption curves for a large number of dyes; absorption curves may also be found in the Colour Index, which may be consulted, on subscription, *via* the Internet. Note that the scaling on the axes in Gurr's book is different from that used by Baker.

Modification of the auxochrome groups will change the binding characteristics of the dye. Acid fuchsine, for example, has three sulphonic residues and two amine residues; it thus has a net negative charge and is thus an anionic dye. Basic fuchsine is a basic dye.

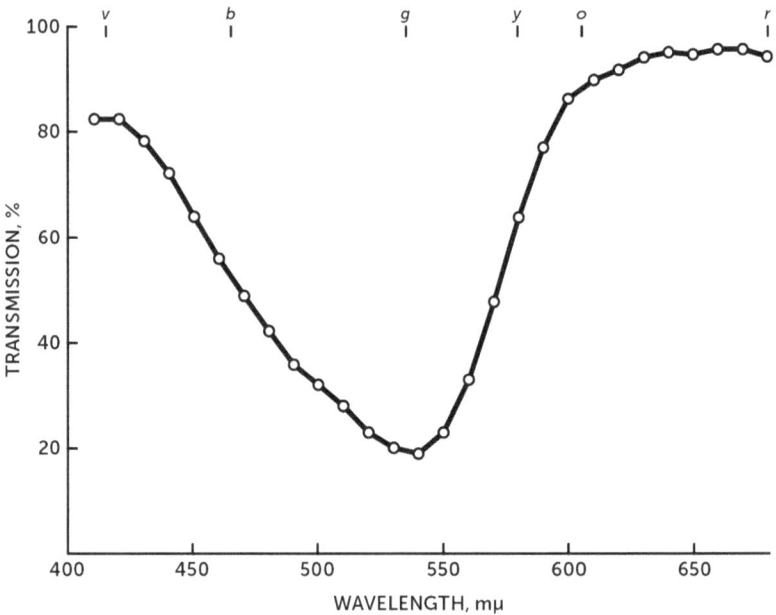

Figure 8.1   **Transmission curve of basic fuchsine.** v = violet; b = blue; g = green;
y = yellow; o = orange; r = red. Note that "mµ" as the units for wavelength is the old nomenclature for nm.

Ionisation of the auxochrome groups depends on pH: we shall come
to this shortly. Modification of the auxochrome groups, for example,
by replacing hydrogens with methyl groups in the amine residue, will
change the colour of the dye.

   Modifications at other positions in the dye molecule can also
affect the colour of the dye. Rosaniline has a methyl group
instead of a hydrogen on one of the aryl rings: it is bluer than
pararosaniline.

8.2.3   Naming of Dyes

The naming of dyes lacks scientific method. Of course, for any dye, a
proper chemical name exists but is usually too long to be remembered
or used every day. Each dye also has a Colour Index (CI) number, but
unless the CI list is to hand this, too, is not memorable. Common
names for dyes are generally used; some make sense, others do not.
The following examples of the reasons for naming certain dyes are
illustrative of this point.

Named for colour	Light green
Named by colour association	Dahlia (dahlias come in many colours; the dye is a violet colour), eosin: the colour of the dawn sky, Sky blue *etc.*
Named by association with an event	Congo red, discovered just as the Congo Free State was established (hardly a self-explanatory name!)
Named by "false chemistry"	Methyl blue has no methyl group; azo carmine has no azo group and is not chemically related to carmine

To make matters worse, any one dye might be known by several common names: "Light Green SF Yellowish" is also known as: Light Green, Acid Green, Lissamine Green SF, Acid Green 5, Food Green 2, FD&C Green no. 2, Green no. 205, Acid Brilliant Green 5, Pencil Green SF, or CI 42095. The only sensible thing to do is to look at the Colour Index, use the common name recommended there and quote the CI number. CI numbers change from time to time: the current edition of the Colour Index is the fourth. Gurr, 1960, tried to get around this by introducing his own "Michrome numbers" (Light Green SF Yellowish: Michrome number 240), which he intended to hold stable, but these are now obsolete. Despite the obsolete numbering, Gurr's book remains a valuable source of reference and contains information not easily available elsewhere.

8.2.4 Chemical Classifications of Dyes

Baker divided dyes into quinonoid dyes, azo dyes and dyes containing a nitro group. Table 8.2 provides examples of each type and a subclassification of quinonoid dyes. I have adhered to Baker's spellings.

This classification may be modified by separating the synthetic dyes from the natural dyes. Conn adopted this approach and included haematein and haematoxylin, orcein, litmus and brazilin amongst the natural dyes.

The key chemical groups that define the groups of dyes listed in Table 8.2 are shown in Scheme 8.7.

8.2.5 Mechanisms of Attachment of Dyes to Components of Tissues

Staining, as understood by histologists, is mainly a matter of dyeing of protein. Some dyes become attached directly to proteins, others require an intermediate substance or mordant.

Table 8.2   Classification of dyes.

Class	Sub-class	Examples of basic or cationic dyes	Examples of acidic or anionic dyes
Quinonoid	Triarylmethane dyes	Basic fuchsine, crystal violet, dahlia, methyl green, aniline blue	Acid fuchsine, methyl blue, light green
	Haematein		Haematein
	Anthraquinonoid		Alizarine, Kernechtrot, carminic acid
	Xanthene	Pyronine G	Eosin Y, erythrosine B, phloxine, fast violet A2R
	Azine (quinone-imine dyes)		
	1. Oxazine	Brilliant cresyl blue,	
	2. Thiazine	Nile blue A, gallamine	
	3. Azine	blue,	
		Thionine, methylene blue, toluidine blue	
		Neutral red, safranine O, Janus green B	Azocarmine G, nigrosine
Azo	Mono-azo	Janus green B (contains an azine and an azochromophore)	Orange G, Chromotrope 2R, methyl orange
	Disazo	Bismarck brown Y	Congo red, trypan blue
	Trisazo		Chlorazol black E
Nitro			Picric acid

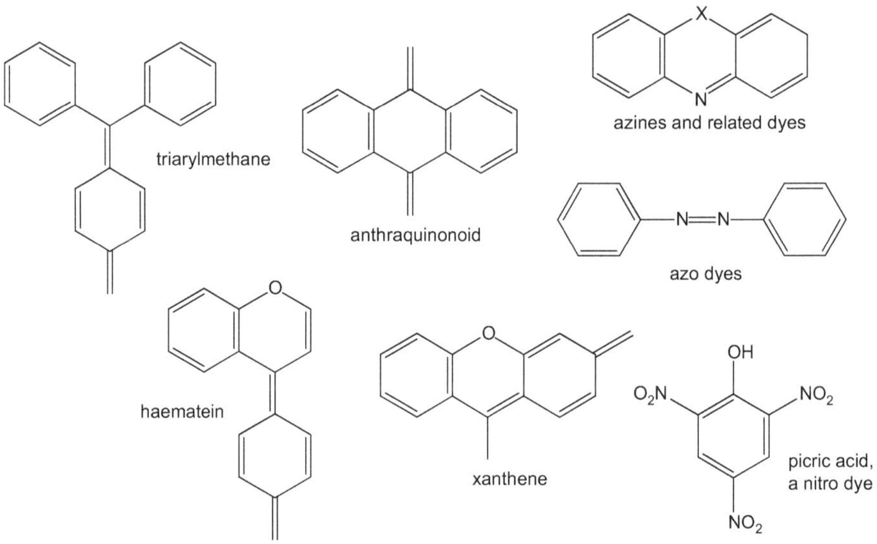

Scheme 8.7   Key chemical groups and main formulae that define the groups of dyes listed in Table 8.2.

8.2.5.1 Direct attachment. Dyes, like proteins, carry chemical groups that can undergo ionisation. The key groups and their ionised forms in dyes are given by Baker as: the amine group ($-NH_2$/$=NH_2^+$), the carboxyl group ($COOH$/COO^-), the sulphonic group ($RHSO_3$/RSO_3^-) and the hydroxyl group (OH/O^-). The amine and carboxyl groups are, of course, also the key ionising groups of amino acids. It comes as little surprise that ionic bonds may be formed between the negative and positive groups of dyes and proteins.

It will be well known that proteins can be characterised by their iso-electric point—that is the pH at which the negatively charged groups of the constituent amino acids are balanced by the positively charged groups. The iso-electric point will depend on the ratio of positive to negative groups. Dyes are smaller molecules than proteins but, with these compounds also, an iso-electric point may be defined. A dye such as basic fuchsine will have a high iso-electric point and at most pH values will be positively charged. Such dyes are described as basic dyes. Orange G, on the other hand, has two sulphonic groups and will be negatively charged at most pH values.

Different components of tissues contain different proteins and at, say, pH 7.0 will give different reactions depending on the preponderance of acid and basic groups. Baker provided a table showing some key components of tissues and their reactions (Table 8.3). Substances that are themselves acidic, attract basic dyes and are described as basophilic; substances that are themselves basic, attract acidic dyes and are described as acidophilic. The term basophilic is widely used but Baker, in an appendix dealing with spelling, which only he would, or perhaps could, have written, explains that basiphilic

Table 8.3 Components of tissues and their staining reactions. Modified from Baker, 1958.

Characteristic	Examples
Acidic materials: basophilic	DNA and chromatin
	RNA and ribonucleoprotein
	Matrix of cartilage
	Many mucous substances
	Most lipids other than triglycerides
Amphoteric	Cytoplasm of most cells
	Contractile substance of muscle
Basic materials: acidophilic	Collagen
	Cytoplasm of red blood cells
	Granules of eosinophil leucocytes
	Nuclei of spermatozoa of certain fishes

is the correct term and that basophilic is not. Indeed, he goes further and points out that basiphil and acidophil are more correct than basophilic and acidophilic. His advice has been widely ignored.

Effects of acids and alkali on direct dyeing

Acid solutions suppress ionisation of acid groups of proteins. Thus, staining with a basic dye (which binds to the acid groups of the protein) will be suppressed by acid solutions. This is what happens when a section stained with haematoxylin is differentiated in acid alcohol. Alcohol alone acts as a weak acid and will also remove basic dyes from a section. Alkaline solutions suppress the ionisation of basic groups of protein. Thus, staining with an acidic dye will be suppressed by alkaline solutions and dyeing with acid dyes will be reduced by exposing tissues to a high pH. This is why tap water washes out the eosin from sections. Rapid dehydration to alcohol reduces the loss of eosin from the sections.

The attentive reader will now be wondering: if the pH of the solution can affect the ionisation of the groups on the protein, can it not also affect the ionisation of the dyes? Baker explained that most dyes remain cationic or anionic throughout the range of pH values used in staining. This is some relief: we need think only about the effects of pH on the cationic and anionic groups of the protein.

Effects of fixatives on direct dyeing

Formaldehyde and mercuric chloride cross link proteins by binding to basic amine groups. It is hardly surprising, then, that these fixatives favour staining with basic dyes. The acidic groups of the protein are unaffected by the fixative. Staining with acid dyes is favoured by trichloracetic acid, picric acid and chromium trioxide. The obvious, but perhaps incomplete, explanation is that the acid groups of protein are suppressed by these fixatives and that the basic groups of the protein are left unaffected. Some fixatives seem to enhance the staining of most components of tissues. Mercuric chloride is the best example of this effect but the mechanism underlying the effect is very imperfectly understood. Baker ventured that mercuric chloride fixed tissues in such a way that they became more permeable to dyes.

8.2.5.2   Indirect Attachment: Mordants.   The word mordant is derived from the French *mordre*, meaning to bite. The word was used in textile dyeing to suggest that the mordant caused the dye to bite into the fabric. A mordant is a chemical compound that forms a link between the dye and the substance to be dyed. Not many dyes are used with mordants in histological work: haematoxylin and carmine are the two major examples. When a mordant binds to a dye, a lake is formed. It would be difficult to think of a less self-explanatory term than "lake" as used in this context. Baker explains, once again as only he could, that "lake" comes from the Hindustani word *lac*, which in turn means the waxy secretion of certain pigmented insects. The waxy substance was called *lac* (hence shel*lac* when the wax is sold in plates), but the word came to be used for the dye extracted from these *lac* insects. The extraction of the dye involved the use of potassium alum, which formed an insoluble compound with the dye. The product was called a lake. Cochineal is the name given to dried female cochineal beetles (not the same as *lac* insects): the dye, carminic acid, is also extracted with potassium alum. The precipitate is called carmine or crimson lake. Haematoxylin lakes are invaluable to the histologist because they are "fast", *i.e.*, resistant to removal by water and alcohol. Once chromatin has been stained with a haematoxylin lake, the remaining components of the tissue can be stained with aqueous or alcoholic solutions of other dyes without de-staining the chromatin.

Mordants can cause changes in the colours of dyes. Haematein, the oxidation product of colourless haematoxylin, is an unimpressive dye of a reddish colour at pH 7 and yellow at pH 1. Once combined with an appropriate mordant, the dye becomes blue or blue–black. Ehrlich's haematoxylin contains haematein (the haematoxylin has been oxidised) and acetic acid: the solution is dark red. The redness is caused by the acetic acid preventing the haematein combining with the alum that is present. Dilute acid is used for de-staining sections that have been stained with alum haematoxylin. It will be noticed that the nuclei of cells become red, and have to be blued by exposure to alkaline tap water. Different mordants produce different degrees of fastness of haematoxylin to acids: iron haematoxylin is much less easy to remove with acid than haematoxylin made up with, for example, potassium alum.

Alums are the great mordants, at least from the histologist's point of view. "Alum" is derived from the French *alute* meaning tawing. Tawing is a process akin to tanning and is, or was, used in the preparation of leather: alum was used in the process. Alums are double salts: aluminium is often one of the cations present, but ferric alum contains iron and chrome alum contains chromium. The second cation is usually potassium, ammonium or sodium. One might have thought that "alum" was derived from "aluminium": seemingly not. All alums crystallise with 24 molecules of water in the alum molecule. The alums that concern us are:

Potassium alum	$Al_2(SO_4)_3 \cdot K_2SO_4 \cdot 24H_2O$
Ammonium alum	$Al_2(SO_4)_3 \cdot (NH_4)_2SO_4 \cdot 24H_2O$
Iron alum	$Fe_2(SO_4)_3 \cdot (NH_4)_2SO_4 \cdot 24H_2O$
Chrome alum	$Cr_2(SO_4)_3 \cdot K_2SO_4 \cdot 24H_2O$

The cations of alums exist in crystals and in solution as complex ions. The action of chrome alum, for example, is $[Cr(OH_2)_6]^{3+}$ in crystals and the crystals are black–violet in colour. In solution the cationic from is $[Cr(OH_2)_5(OH)]^{2+}$ and the solution is green. These complex cations are chelated by dye molecules. Chelated? Of course, from *chela*, meaning claw, indicating two bonds like claws or pincers! One of the bonds, referred to as the primary bond, is covalent; the secondary bond is a dative bond with all the necessary electrons being provided by a conveniently placed oxygen atom. The primary bond is formed between a phenolic OH group and the metal.

The mordant is positively charged and forms ionic bonds with negatively charged groups of molecules within tissue. Thus, for our purposes, all mordant dyes are basic dyes and stain basophilic substances, like chromatin. Might it be possible to have a negatively charged mordant? Yes, tannic acid is used in the dye industry in this way. But in histology we need think only of cationic mordants.

Tissue stained with a mordant–dye complex (a lake) may be de-stained in three ways:

1. With the mordant itself: an equilibrium between the dye attaching to the free mordant and to the mordant that has become attached to the tissue is established and, if more free mordant is provided, the dye will leave the tissue–mordant complex.
2. With acid: acid will suppress the formation of negatively charged groups on the protein molecules.
3. With oxidising agents, such a potassium permanganate.

8.2.6 Digression on Leucobases (for Enthusiasts Only)

Many dyes can be converted to colourless compounds by the action of reducing agents on quinonoid linkages. The colourless compounds formed are described as leucobases. The formula of

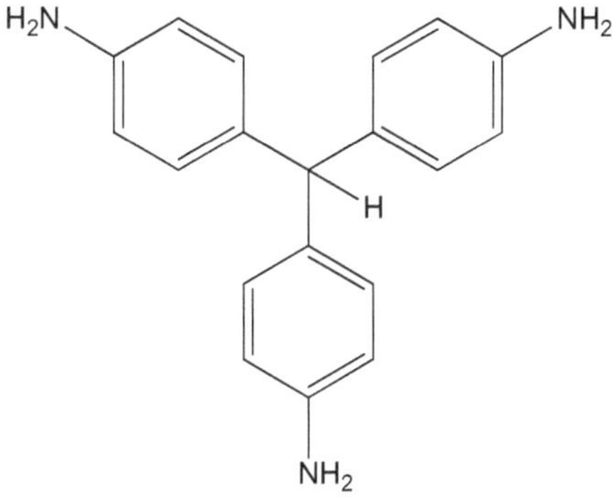

Scheme 8.8 Chemical structure of leuco-*para*-rosaniline.

leuco-*para*-rosaniline is shown in Scheme 8.8. Leucobases become coloured on oxidation.

Dyes can also be oxidised to colourless compounds: potassium permanganate and potassium ferricyanide work in this way with dyes used with mordants. Schiff's reagent is a decolourised dye but it is not a leucobase: it becomes coloured after reaction with an aldehyde group but this is not as a result of oxidation.

8.2.7 Accentuators and Accelerators

A number of compounds have been found to help the process of dyeing but not by acting as mordants. Carleton referred to these as accentuators and accelerators and provided the following list. How these compounds work is not quite known.

8.2.7.1 *Accentuators*

- Potassium hydroxide (included in Loeffler's methylene blue method);
- Borax (sodium tetraborate: again used with methylene blue);
- Aniline (used with a number of dyes, including safranin and gentian violet).

8.2.7.2 *Accelerators.* These are used to enhance metallic impregnation methods used for staining the central nervous system. Rather interestingly, they are largely hypnotic substances.

- Chloral hydrate;
- Veronal (barbituric acid, barbitone, 5,5-diethylpyrimidine-2,4,6(1*H*,3*H*,5*H*)-trione);

- Sulphonal (acetone diethyl sulphone);
- Trional (2,2-bis(ethylsulfonyl)butane);
- Hedonal (pentan-2-yl carbamate);
- Nicotine.

Glycerol is included in some staining mixtures in order to slow the evaporation of the stain from slides when the staining on glass rods method is used. Carleton commented that glycerol tended to slow down staining processes.

8.2.8 Differential Staining

Histologists stain tissues so as to produce contrast: contrast of colour between different components of the tissue. Perhaps the most widely used examples of such stains are the trichrome stains used for connective tissue. More than three stains can be used: a trichrome plus a stain for elastic fibres plus a silver impregnation method for reticulin fibres—five or six different stains (using the term loosely to include impregnation) are in play at the same time. Unsurprisingly, as more colours are added, the section requires more careful differentiation to prevent a muddy appearance similar to that produced by mixing colours from a paint box. There is much to be said for taking serial sections and staining each for an individual component of tissue. The standard H&E technique is, of course, a differential method: the basic material of the cytoplasm stains with the acid dye eosin and is described as acidophilic; the acidic material of the chromatin stains with the basic dye (see above).

Differential staining depends on two things: the physico-chemical characteristics of the components of the tissue and the physico-chemical characteristics of the dye:

- Tissue: preponderance of basic or acidic groups at the pH used in the staining bath (see above), density in terms of number of protein molecules per unit volume of tissue and permeability in terms of how easy it is for dye molecules to penetrate into the component.
- Dye: acid or basic groups at the pH used in the staining bath (see above), molecular weight, which controls the rate at which the dye will diffuse, and whether the dye exists as a solution containing individual molecules or whether the molecules aggregate or whether the dye exists as a colloidal system.

The permeability of tissues and tissue components contains some surprises: collagen is much more permeable to dyes than red blood cells. One might have thought that collagen was a dense material (it is) and difficult for a dye to penetrate. This is not the case. Chromosomes are dense structures in comparison with "ordinary" cytoplasm and stain both darkly and easily. Elastic fibres are dense and, though orcein does not bind particularly strongly to elastic fibres in comparison with other tissue components, washing out of the dye with acid alcohol leaves the elastic fibres darkly stained.

Baker pointed out that phosphomolybdic acid was often referred to as a mordant for collagen stains such as aniline blue. He added that this was quite incorrect. Recall that the mordants used in histology are basic substances (the alums, for example) and that phosphomolybdic acid is, of course, an acid. What actually happens is that phosphomolybdic acid blocks the dyeing of collagen with one of the red dyes used in the trichrome methods (azocarmine or acid fuchsine, for example) and, after being washed out, leaves the collagen accessible to the blue dye, aniline blue. Washing in phosphomolybdic acid will remove the red dye from the collagen before removing it from cytoplasm and muscle fibres. Aniline blue actually diffuses more slowly than the red dyes, and if placed in direct competition, so to speak, with the red dye would not stain collagen at all well: the collagen would be stained by the red dye. Baker pointed out that phosphomolybdic acid was really acting as a competitive "colourless dye". Far from mordanting aniline blue, it actually opposes the action of aniline blue: it prevents it staining the cytoplasm and protects the collagen from the red dye. If a section that has been nicely stained for collagen with aniline blue is washed in phosphomolybdic acid, the aniline blue will be removed from the collagen.

The rate at which a dye diffuses depends on its physical state and this can be affected by temperature. Mitochondria cannot be stained with acid fuchsine at room temperature but stain well close to the boiling point of water. What has happened is that the molecules of the dye that are aggregated at room temperature have been separated from one another at higher temperature and have been able to diffuse into the mitochondria. Once the temperature is lowered, they do not diffuse out of these organelles and another dye that could enter the mitochondria at room temperature will find the available binding sites taken. Baker pointed out that red blood cells that do not take up acid fuchsine at room temperature are well-stained at higher temperatures: again, the dye molecules have been separated and have diffused into the rather impermeable red cells.

It is fair to say that these principles that control the differential effects of mixtures of dyes have usually been worked out after the properties of the mixture have been discovered: the development of staining techniques has relied on empiricism rather than on theory.

8.2.9   Metachromasia

To the histologist, metachromasia means a change in the colour of a dye when it becomes attached to a component of tissue. For example, toluidine blue is a blue dye and stains some components of tissues blue. This is called orthochromatic staining. But the matrix of cartilage is stained a purple–red: this is an example of metachromatic staining. Components of tissues that cause a dye to stain metachromatically are described as chromotropes. Histologists distinguish between metachromasia and the effects produced by dyes that contain impurities or that have undergone partial conversion to other dyes: although these may produce effects that look like metachromasia, they are referred to as allochromasia. This subject is bedevilled by long words! Typical examples of metachromatic dyes include methyl violet, thionine and toluidine blue. Metachromatic dyes are not confined to any one of the groups of dyes described above: methyl violet is a triarylmethane dye, toluidine blue is a thiazine dye.

The effects of chromotropes on the colours of metachromatic dyes are not easy to explain. Consider a purple dye. The dye transmits light from both ends of the visible spectrum but absorbs light from the middle of the range. Thus, red and blue/violet light is transmitted and the dye looks purple. If the absorption curve were to be shifted towards the violet end of the spectrum then more red and less blue light would be transmitted. The dye would change in colour from purple to a more reddish colour, perhaps magenta. If the curve was further displaced in the same direction, no blue or violet light would be transmitted but a lot of red and orange, and perhaps yellow, light would come through: the dye would appear reddish-orange. The sequence: purple, red, orange, yellow describes what happens and this is also described as a hypsochromic effect. This is the usual form of metachromasia seen in histological work. The opposite effect is possible: this is described as a bathochromic effect, but we shall ignore it except to say that it seems to be a property of some acidic dyes. Let us consider hypsochromic effects a little further. Imagine that the dye began, so to speak, as green in colour. Green light is transmitted and blue and red light is absorbed. Push the curve to the left: even less red, orange and yellow will be transmitted, but more blue will be transmitted. The colour change is from green to blue. Similarly, if the dye began as blue, it would become violet. The full sequence of hypsochromic colour changes is this: green to blue, blue to violet (or purple if some red light is transmitted from the other end of the spectrum), purple to magenta, magenta to red, red to orange, orange to yellow, and yellow to green. In practice only the purple-to-magenta effects are observed because of the stains usually used. The reason for all this being complicated is that the transmission curve, or absorption curve if you prefer, is usually narrower than the visible spectrum and so effects from both ends of the spectrum have to be considered.

All the metachromatic dyes in general use are basic dyes. This means that all the chromotropes are acidic and we should expect to find them in, for example, tissue components that contain a lot of sulphate groups. Indeed, this is the case: the heparin granules of mast cells and the ground substance of cartilage are classic chromotropes. The presence of metachromatic effects provides insight into the composition of the material involved.

Rather oddly, the exact mechanism underlying metachromasia seems to be unknown. It seems that only when the acidic groups are densely placed, *i.e.*, close together, does metachromasia occur. This suggests that binding of the dye in what Baker called "polymeric form" is important.

8.3 COLOURING BUT NOT DYEING

Most of this chapter has been about the use of dyes; we now turn to three other methods for producing colours (in one case, black—not really a colour) in histological sections.

8.3.1 Silver Impregnation

Silver impregnation techniques are used to "stain" (Baker pointed out that the word "stain" was used to cover such a range of methods that it was essentially meaningless) reticulin fibres, nerve fibres and endocrine cells, for example, those found in the mucosa of the gut. The process is based on that used in photography. A soluble silver salt is reduced to metallic silver by a component of the tissue. In most cases not enough silver is produced to be seen, but when a reducing agent is added to the system further deposition of silver occurs around the original nuclei of deposition. The tissue that has been exposed to silver, and in which only a little reduction has occurred, is in the same state as a photographic plate before the developer is applied: it holds a latent image that can be developed into a visible image. An exception to the need for an added reducing agent (or a developer, to use a photographic term) is provided by the argentaffin cells of the gut. These cells reduce enough silver to be seen without adding a reducing agent. If a section of gut is stained, using the word widely, for argentaffin cells and the next section stained in much the same way, but with a reducing agent included in the process, then a further group of cells will be stained. These are the argyrophil cells. In these cells not enough silver is deposited to allow the cells to be seen, which is a result of the action of reducing groups present in the tissue. Of course, the process for impregnating argyrophil cells will also impregnate the argentaffin cells, and thus two sections are needed to allow these two types of cell to be differentiated. A number of reducing agents can be used, including formaldehyde and hydroquinone, the latter being used in photography. Just as in photography, the development of the picture is stopped by washing in thiosulphate: the photographic "fixer". Silver impregnation cannot be regarded as a histochemical technique because it is not specific for any particular chemical grouping: all it requires is that groups capable of reducing the silver solution are present. Sulphydryl groups play a large part in this but other groups are also involved.

8.3.2 Production of Coloured Material in Tissue

Ferric iron combines with potassium ferrocyanide to produce Prussian Blue. This is used as a histochemical method for detecting ferric iron in sections of tissue. The method is specific for ferric ions.

Another truly histochemical method is provided by the use of Schiff's reagent for staining aldehyde groups. Schiff's reagent is produced by bubbling sulphur dioxide through a solution of basic fuchsine. A colourless compound is formed. This is not a leucobase: leucobases are produced by reduction and are recoloured, so to speak, on oxidation. The leucobase of pararosaniline and the compound that acts as Schiff's reagent are shown in Scheme 8.9.

When Schiff's reagent reacts with an aldehyde the coloured compound shown in Scheme 8.10 is produced. The coloured substance is a dye.

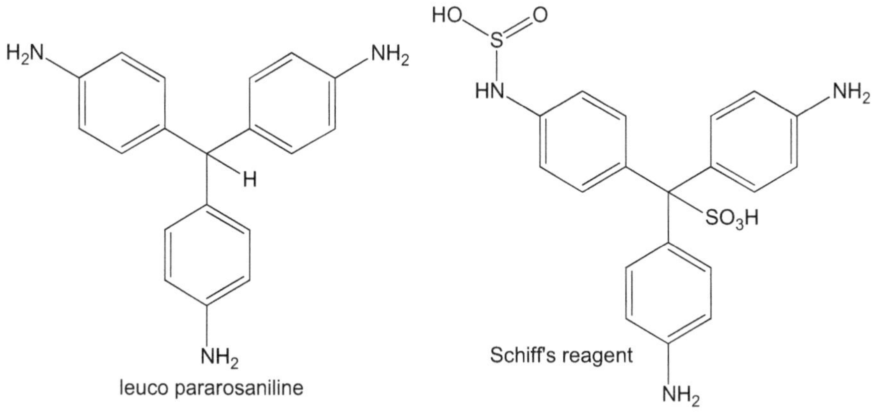

Scheme 8.9 The chemical structures of the leucobase of pararosaniline and the compound that acts as Schiff's reagent.

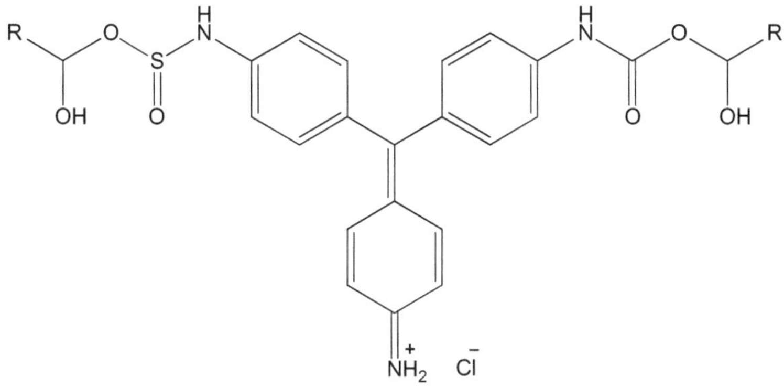

Scheme 8.10 The dye produced when Schiff's reagent reacts with an aldehyde.

8.3.3 Colouring by Dissolving a Coloured Substance in a Fluid Contained in the Tissue

The best example here is the use of coloured compounds, such as Sudan Black or Oil Red O, as "stains" for fats. The formula of Sudan Black is shown in Scheme 8.11. Note that it looks much like a diazo dye but that there are no auxochrome groups present and no ionic bonding to components of tissue occurs. The requirement here is that the coloured substance should be soluble in fat. It must be presented to the fat in a form that will allow it to diffuse into the fat. In chemical terms the partition coefficient between the fat and the vehicle in which the colour is dissolved must favour the fat. Ethanol provides a

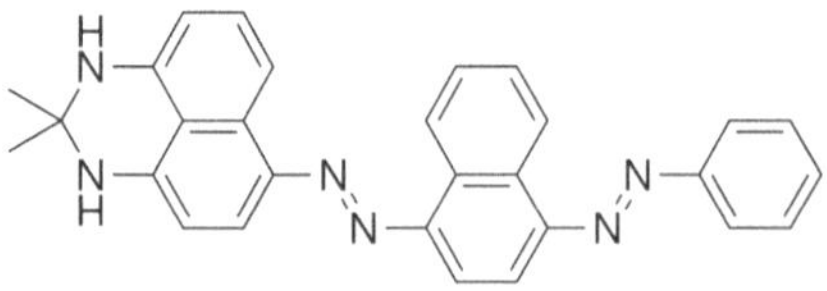

Scheme 8.11   Chemical structure of Sudan Black.

suitable vehicle. Substances such as chloroform or xylene would dissolve the coloured material but would also dissolve the fat. Coloured materials used in this way are described as lysochromes. Naturally occurring lysochromes include the red carotene material that can be extracted from red peppers.

8.4   CONCLUSION

It is not always necessary to understand the chemistry behind the staining reactions that you use. However, it can be of great help in determining if a particular stain is of use to your study. Or if, rather unfortunately, your staining has not worked or disappears upon mounting your section, an understanding of the chemical properties can help you set it right. Hopefully, this section has provided you with at least a frame of reference for this, even if you pick a stain and never think about the chemistry again! These principles will stand you in good stead as you move into the processes of histochemistry set out in the next chapter.

CHAPTER 9

Histochemistry

The processes of histochemistry and its sub-type, immunohistochem-istry (IHC), allow the researcher to delve further and more specific-ally into the chemical characteristics of their specimens and samples. They can specifically identify proteins and reveal changes that have been induced by your experimental manipulations. Para-mount to producing standardised and reproducible results is the ini-tial design and preliminary optimisation of the histochemical study. Therefore, clear positive and negative controls are essential for deter-mining the success of your staining whether this is using chemical or immunological interactions for identifying your target of interest.

There are many different histochemical techniques available; we focus on the most common and standard methods for specifically labelling proteins in formalin-fixed paraffin tissue sections. In par-ticular, we will devote the latter half of the chapter to the discussion of immunohistochemistry, which has become a vital component of re-search studies across many disciplines. Due to the occasionally fin-icky nature of IHC optimisation, we have also provided some basic advice on troubleshooting your protocols.

9.1 PRINCIPLES OF HISTOCHEMICAL METHODS

Though one of the earliest uses of histochemical techniques, the identification of starch by use of iodine predates the development of histological techniques and histochemistry is usually seen as a

Histological Techniques: An Introduction for Beginners in Toxicology
By Robert Maynard, Noel Downes and Brenda Finney
© Maynard, Downes and Finney 2020
Published by the Royal Society of Chemistry, www.rsc.org

refinement of "ordinary" histological methods. Histochemistry is perhaps best described as analytical chemistry on the section: the emphasis is, again, on the mechanism and on the precision of the techniques. The essence of histochemistry is the use of methods that lead to specific reactions with components of tissues and cells and that generate visible reaction products. To be useful, a histochemical technique has to produce something that can be seen; ideally, it should produce something that can be measured. In this, the methods differ not very much from those of histology. Elastin, collagen and nuclei can be stained specifically, and the depth of staining assessed. Some methods usually regarded as histological are, in fact, histochemical: staining for iron, for amyloid, for melanin and for carbohydrates by creating aldehyde groups and then staining for them are all histochemical methods.

The distinction between histological and histochemical methods is useful in practice but by no means absolute. Many of the standard stains described in Chapter 7 allow for visualisation of specific components and are histochemical techniques. Histochemical techniques focus on chemical components of tissues and cells and can be used, very importantly, to measure and locate these components. The components may be naturally occurring, for example, the enzyme alkaline phosphatase, or abnormal as, for example, in the case of the increased deposition of iron in cases of haemosiderosis. Immunohistochemistry (IHC) is also a histochemical method as the interaction and binding of antibody to its target is a chemical reaction. In many cases for visualisation of this reaction, what has been done is to combine the technique of enzyme-histochemistry with that of IHC. Immunohistochemistry is often referred to as IHC, but immunofluorescence (IF) and immunocytochemistry are other terms used in this area.

Gomori (see Bibliography) divided such histochemical methods as were available in 1952 into those for identification of:

1. Inorganic substances, including metallic and non-metallic elements;
2. Organic substances, including saccharides (including nucleic acids), lipids, proteins, amino acids and products of protein metabolism, prosthetic groups, various unclassified organic substances and pigments;
3. Enzymes.

This simple classification is still useful. Of course, the number of methods available has increased during the past seventy years! The principles underlying the methods have not, however, changed very much.

All histochemical methods are designed to identify and allow the location of chemical components in tissues and cells. This is accomplished by causing the components to react with reagents so as to produce something that is coloured and that can be seen with the microscope. Often, a coloured reaction product is produced; sometimes, the reaction leads to binding of a visible marker to the target compound. If T represents the target and R the reagent then, at its simplest, the reaction can be shown as:

$$T + R \rightarrow V, \tag{9.1}$$

where V is the visible product.

The reaction can be complicated by adding further steps:

$$T + R_1 \rightarrow V_1 \tag{9.2}$$

$$V_1 + R_2 \rightarrow V_2. \tag{9.3}$$

V_2 might be visible, whereas V_1 might not be.

Let us look at a yet more complicated example. Let T be a target protein and Ab an antibody raised against T. Let the antibody be linked to an enzyme E to produce Ab−E. The enzyme E catalyses the reaction:

$$R \rightarrow V, \tag{9.4}$$

where V is visible.

Then:

$$T + Ab{-}E \rightarrow TAb{-}E. \tag{9.5}$$

$$TAb{-}E + R \rightarrow TAb{-}E + V \tag{9.6}$$

As long as V remains visible and collocated with T, we will have found a method for identifying T and locating it in tissues or cells.

It will be obvious that a good deal could go wrong in what can be a complicated process. The antibody might fail to bind to the target or perhaps it will be found to bind to things other than the target; perhaps the visible reaction product will diffuse away from the target or break down and disappear. Perhaps the serum containing the antibody will contain other antibodies and proteins that will bind to components of the tissue or cell other than the target. All these things could go wrong; therefore, very careful technique and the astute employment of controls will be necessary if the target is to be identified and located with confidence. It is in this that histochemical techniques differ from histological techniques: more care is needed to ensure that the correct conclusions can be drawn.

9.2   REQUIREMENTS OF HISTOCHEMICAL TECHNIQUES

The essential requirements of a histochemical technique are:

1. It must be specific for the target;
2. It must be sensitive and allow identification of small amounts of the target;
3. It must allow accurate localisation of the target.

These requirements need some thought to address, not only with the histochemical method but also, and very importantly, how the tissue has been processed before sections are exposed to the technique. It will be clear that the target must not have been destroyed during the processing steps. It will also be clear that the target should not have been allowed to move in the tissue, or cell, during processing. Can this happen? Yes, indeed: glycogen is often displaced by diffusion of fixatives: a phenomenon charmingly known as "glycogen flight". In many cases the type of identification desired will determine the processing treatment of your samples. However, it is also possible to work backwards to an appropriate histochemical method if you have archived samples, as long as you know how they were processed initially.

Gomori lists further requirements:

1. Reactions must take place *in situ*;
2. Tissue structure must be preserved;
3. The end-product must be reasonably stable and certainly not transient;
4. Crystals, if produced, must be small enough to allow localisation in the tissue or cells;
5. Reactions should be rapid: slow reactions are likely to provide time for diffusion of reactants and products.

The beginner might, by now, be thinking that histochemistry is a demanding subject. It is, but 60 or 70 years of research and development have ironed out many of the difficulties and, today, histochemical techniques can be used almost, if not quite, as easily as ordinary histological methods.

9.2.1 Controls

The use of control sections is central to histochemical work and deserves some discussion. We can think in terms of positive and negative controls. Positive controls allow you to test your method, negative controls test the specificity of the method.

Is your method capable of detecting and localising the target? You might think that, if you follow an established and well-tested method, an apparent failure to demonstrate the presence of the target would mean that the target was not present in your tissue samples. This is not unreasonable, though a failure in selective staining of nuclei with haematoxylin would be unlikely to make you think that you had struck a rare tissue in which the cells lacked nuclei. You would be more likely to think that something had gone wrong with the method. The difficulty in histochemical work is that you might well not know whether the target is present or not. It might have been destroyed or, in the case of an enzyme, inactivated during processing or by the chemical to which the experimental animal was exposed. Running positive controls resolves this difficulty.

Sections for use as positive controls may be taken from tissue collected from animals not exposed to the chemical being studied or from a library of tissue that has been built up and studied over the years. Sections being used as positive controls should be passed through exactly the same steps as the test sections. This is vital. It is obviously useless to look at a five-year-old section in which, for example, alkaline phosphatase was clearly demonstrated and to use this as a basis for judging the success of your work today. The method might have changed or, and importantly, you might have made a mistake. The golden rule should: **always be obeyed**.

Imagine that two chemical species, A and B, are collocated in the tissue you are studying. Your method is designed to detect A, but might in fact be detecting B. How could you know that this was the case? The answer is by blocking the reaction with A and demonstrating that no reaction occurs. This leads to a difficulty. How can you know that blocking A does not also block B? Only by studying the reactions of the blocking agent with both A and B. Yes, but what if you don't know the identity of B? That is a problem! Fortunately, this difficulty seems to arise infrequently. Another type of negative control is produced by "running a blank", meaning omitting an essential step in the histochemical method. For example, if D is required for the development of a coloured reaction product then omitting D should lead to an unstained section. If staining does occur then something more complicated is going on and needs investigation.

9.3   WHY USE HISTOCHEMICAL TECHNIQUES?

Before turning to a selection of individual techniques, it might be helpful to consider why a toxicologist might wish to use such methods. The great advantage of histochemical techniques is, as already stated, their specificity. Ordinary histological techniques allow the identification of components of tissues and cells on what might be called a large scale. Haematoxylin stains nuclei but also stains other things. This is because the mechanism of staining relies on ionic bonds: the stain is not, so to speak, fussy about what molecules provide the appropriately charged bonding sites. Histochemical techniques are, on the other hand, very fussy about the molecules with which they react or to which they bind. If you are interested in identifying specific molecules then you will need to use histochemical techniques. This suggests several possibilities to the toxicologist. You might be looking for metals, iron, mercury, cadmium, tin or for arsenic. Or you might be looking for a loss of glycogen, for an increase in lipid or for the appearance of amyloid. You might think from clinical observations that the compound you are studying inhibits an enzyme and want to look at the amount of that enzyme present in specific cells. For all these you need to use histochemical techniques.

Toxicology is a science and the essence of all sciences is measurement. Histochemical methods allow measurement of the amount of, or the activity of, the target in tissues. Perhaps the easiest measurement method to understand in principle is the quantification of cells in a tissue that are undergoing proliferation. You can use a label that detects proliferating cells in your tissue of interest and count the number of these cells, as well as the number of non-proliferating cells. You then count, say, 1000 cells per tissue in both your experimental samples and your controls. The resulting numbers would allow you to determine if there was a relative difference in the amount of proliferation going on in the two groups. Books listed in the Bibliography provide details of such advanced methods. To sum up: use histochemical methods to find the target, to define the distribution or location of the target and to measure how much of the target is present and how it is distributed in quantitative terms.

Examples of histochemical protocols for staining of alkaline phosphatase are described in the following sections.

9.4 ALKALINE PHOSPHATASE HISTOCHEMISTRY

There are several different techniques for demonstrating alkaline phosphatase activity, the highest levels of which are found in the liver and in bone, with lower amounts in the small intestine. The three methods excellently described by J A Kiernan (see Bibliography) are: cerium-DAB, azo-coupling and indoxyl-tetrazolium. These methods require specific fixation methods in order to preserve the pattern of expression. For example, the cerium-DAB method requires vascular perfusion of cacodylate-buffered gluteraldehyde and sectioning on a cryostat, while the azo-coupling method recommends fixation in cold acetone or formalin with careful paraffin embedding. Each of these methods uses a different substrate to visualise the alkaline phosphatase: diaminobenzidine shows activity as blue–black, α-napthyl phosphate shows as purple to black and 5-bromo-4-chloroindoxyl phosphate results in a blue product. Ultimately, the method chosen will depend on the capabilities of your lab for processing your samples, or the availability of the samples themselves. We recommend thorough background research as to the standards of your field and the protocols going on around you as you begin setting up any histochemical protocol.

9.5 IMMUNOHISTOCHEMISTRY

9.5.1 What is Immunohistochemistry?

As stated earlier, immunohistochemistry is a form of histochemistry that uses the reaction of antibodies to their targets for visualisation in samples. It is not surprising to learn that antibodies raised against specific proteins bind to those target proteins when applied to tissues that contain them. The binding is, of course, chemical. Unfortunately, the antibodies may also react with other components of the tissue and an undesirable lack of specificity may be noticed. Dealing with background staining, or off-target binding, is one of the arts of IHC.

We cannot, of course, see antibodies in sections, even with a microscope, without making them stand out in some way. Figure 9.1 demonstrates these concepts schematically. This can be done by labelling the primary antibody that becomes attached to the target protein with a chemical that fluoresces when exposed, for example, to

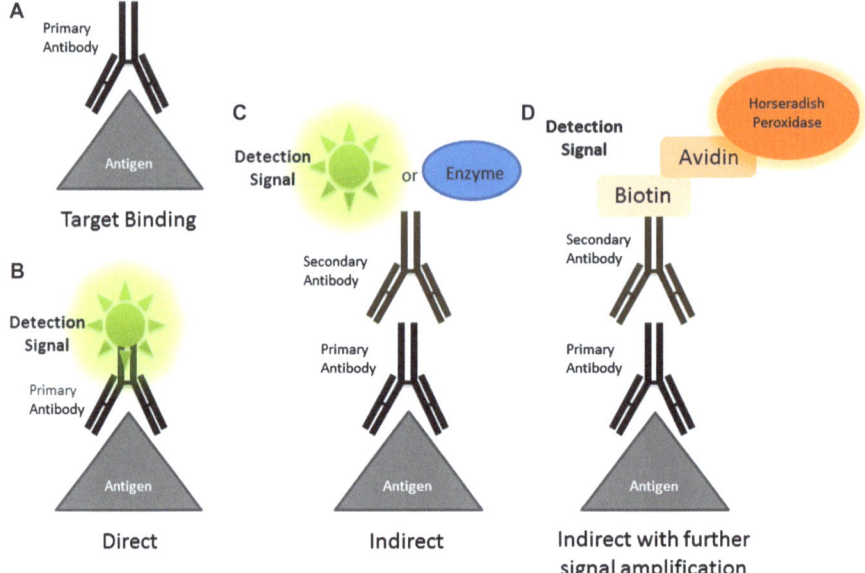

Figure 9.1 **Basic principles of immunostaining. A.** A primary antibody recognises an antigenic epitope on a target protein. **B.** Direct immunofluorescence detection is accomplished by using a primary antibody to which a fluorescent molecule has been directly conjugated. **C.** Indirect immunodetection is where a secondary antibody is applied to the sample; this secondary antibody could be conjugated to a fluorescent label or an enzymatic tag. While this schematic shows a 1 : 1 binding of the secondary antibody to the primary antibody, in practice several secondary antibody molecules can bind to a single primary molecule. **D.** Indirect immunodetection with signal amplification can be accomplished by several different methods. The one represented here is using a biotin–avidin–horseradish peroxidase reaction, where the secondary antibody is biotinylated; application of the avidin–horseradish peroxidase protein conjugate further amplifies the signal before colour development with a substrate chromagen.

ultra-violet light. The sites where antibodies have bound will stand out from the non-fluorescent background. Alternatively, the antibodies can be located by attaching another, secondary, antibody (an antibody to an antibody!) that is, in turn, attached to a molecule that is fluorescent or to an enzyme. This sounds complicated; let's go a little further. If the enzyme can still function then it can be used to catalyse some reaction (we shall provide the reactants) and the products of that reaction might be visible or easy to make visible. Thus, the protein we are looking for might be identified by its binding to a primary antibody, which is bound to a secondary antibody, which is, in turn, bound to an enzyme. Use of a secondary antibody is helpful: several secondary antibodies may bind to the primary antibody, and thus the visual signal can be enhanced. Special kits that allow further signal amplification are available: these will be discussed later in section 9.5.7. No immunohistochemical technique can work without antibodies; let us look next at antibody production.

9.5.2 Antibody Production

We shall begin with primary antibodies. These are the antibodies that bind to the target protein. For example, if we are hoping to identify and locate the muscle protein actin in a section, it makes sense to use an antibody raised against actin. On the actin protein there will be several antigenic sequences (a sequence of amino acids), which are called epitopes. Primary antibodies (in practice, the singular "primary antibody" is often used) are classified as either polyclonal or monoclonal. Polyclonal antibody is raised by repeatedly injecting an animal, for example, a rabbit, with the protein we are trying to detect. If the immunisation is successful, the rabbit will produce antibodies and these will be found in serum. This serum will contain antibodies to different epitopes, as well as different antibody isotypes (*e.g.*, immunoglobulin G (IgG) or IgM). When buying a commercially available polyclonal antibody, it has most likely been purified down to a single isotype, but this will still recognise several epitopes.

Monoclonal antibody, on the other hand, is produced by cell culture techniques. A cloned immune cell line, usually combined in a hybridoma, produces just one sort of antibody in response to challenge with a protein. This single-cell-line method of producing antibody cuts down, but may not eliminate, background staining. Monoclonal antibody is often made using mouse hybridomas. Cells from ascites fluid in mice can also be used. Let us imagine that we are looking for a protein in mouse tissue. If our primary antibody was raised in a mouse hybridoma then we might describe our technique as "mouse on mouse". This is likely to produce a lot of background staining and special mouse-on-mouse kits have been developed to reduce this.

Secondary antibody is produced by inoculating animals with purified antibodies (immunoglobulins) from another species. For example, if our primary antibody was raised in a rabbit then rabbit immunoglobulins injected into a goat produce the secondary antibody. Rather obviously, the animal used to produce the secondary antibody must be of a species different from that used to produce the primary antibody. The secondary antibody is not specific to the primary antibody in any way except for the species; therefore, a single secondary antibody can be used to detect several different primary antibodies.

9.5.3 Considerations before Applying IHC Techniques

1. Consider the objective of the study. Are you hoping to identify and locate a single target protein or are you interested in several target proteins? If you are in pursuit of just one protein then using a primary antibody linked to an enzyme would be a suitable technique to choose. If, on the other hand, you are in pursuit of several proteins then using several antibodies, each carrying a different fluorescent marker, would be appropriate. However, it is best to keep things as simple as possible at the beginning. There are numerous courses and extensive manuals if you are looking to use complicated staining combinations and advanced microscopy, a couple of which have been listed in the Bibliography. For the purposes of this chapter, it will be assumed you are going to look at a single target protein using an enzymatic detection method.

2. Consider whether you should buy or make the antibody you need. Antibodies to many proteins are now commercially available and are tested by the manufacturers. Responsible manufacturers provide details of their testing process and links to published works that report use of their product. In this chapter we shall assume that you are using "commercial antibody". When buying primary antibody, it is important to enquire about cross-reactivity between species. If you are looking for a specific protein in both mouse and human tissue, you might need two antibodies. But a rabbit immunised with the mouse form of the protein might produce antibody that reacts quite satisfactorily with the human protein, and thus only one antibody will be needed. Again, this will be listed in the information provided by the manufacturer, if they have tested the cross-reactivity. If an antibody for your protein of interest is not commercially available, it is becoming more and more common to hire a company, or use a core facility, to make the antibody for you rather than jumping in to this at the beginning. If you decide that you are making the antibody yourself, be aware that the optimisation process is very important and can be a significant investment of resources and time.

3. Consider the type of specimen or sample you intend to use. Are you thinking of working with a histological section, cultured cells

or with a whole mount of a piece of tissue? Cultured cells are more delicate and whole mounts are much thicker than sections and special techniques are needed. An example of a whole mount preparation for confocal fluorescent microscopy is shown in Fig. 3B. Generally, tissue sections are a good place to start and, depending on your antibody, you will need to decide whether to use frozen or paraffin sections. Either can be used and the ease of handling paraffin sections makes them attractive. When buying antibody, again look at this information on the data sheet for your antibody of interest. Generally, if you are at the beginning of a project, you choose your antibody or antibodies and then gather and process your samples according to the recommendations published for the antibody. If archived tissue is to be used, you will need to know how it was processed and how it was fixed. In general, most antibodies for IHC will have been optimised for formalin-fixed paraffin sections. However, as stated earlier, some antibodies will not be able to react due to the cross-linking nature of this type of fixative and, therefore, antigen retrieval methods may need to be added to your protocol. Generally, if an antibody works on formalin-fixed paraffin sections then it will also work on frozen sections; however, the reverse is not true. Frozen sections are generally not exposed to a fixative as the quick freezing of the tissue acts as the preservative; therefore, proteins and antigens are in their natural state. If you have an antibody that is highly reliant on the natural conformation of the protein or epitope, frozen sections should preserve this and allow it to be recognised. As in all histological work, the method chosen depends on the objective of the study.

Consider your microscope. If you plan to identify your protein by IHC involving an enzymatic reaction then an ordinary bright-field microscope will be needed. If, on the other hand, you intend to use fluorescence as your marker technique, you will need a "fluorescence microscope". These are widely available, although the technology associated with them is always rapidly developing and changing. With this in mind, a good starting reference can be found in Lichtman and Conchello (2005)[2] and in Chapter 3 on microscopy.

9.5.4 Optimising Your Method (Antibody Optimisation)

Optimising your methods is important in all histological techniques. Even in such a standard method as H&E staining it is important to standardise the durations of exposure of the sections to stain. In IHC techniques this is very important. If, for example, too high a concentration of antibody is used then heavy background staining may be produced and the section will be poorly differentiated as regards to the target protein. As stated earlier, in all histochemical methods controls are essential; however, the type of controls that can be used or produced for immunohistochemistry will vary from those discussed above. Therefore, let's first look at positive controls.

The key features of a positive control are: plenty of target protein and well-understood localisation of the target protein. Many manufacturers list positive control tissues on their websites. If you are preparing the tissues yourself, you should ensure that the positive control and experimental samples have all been prepared in the same way. Of course, you will have to experiment a little: adjusting the dilution of your antibody preparation and the time for which the tissue is exposed to the antibody will be important in producing a properly differentiated positive control.

Negative controls can be produced using the same tissue used for the positive controls. In essence, one can either omit the primary antibody altogether or inhibit the primary antibody by pre-exposing it to an excess of the protein to which it was raised. This sounds a little complicated. Let's consider our target protein P. If we expose the primary antibody serum to an excess of P, we expect all of the individual antibodies in the primary antibody serum to bind tightly to P. The primary serum should now be useless for staining P in tissue sections as all of the binding sites are now full: a negative control will have been produced. It is important when thinking about controls to be clear about what you are actually controlling. Are you producing a negative control by omitting or inhibiting the antibody or by choosing a section without any of the protein to which the antibody was raised? In the former case you are looking at the capacity of the antibody to bind to your target; in the latter you are looking at the capacity of the antibody to bind to things other than the target.

Once you have a good, well-differentiated, positive control and a clear negative control you may congratulate yourself that you have optimised your method.

9.5.5 Antigen Retrieval Methods

As mentioned earlier, the process of fixation can change the ability of a primary antibody to bind to your protein of interest. This could be due to the cross-linking nature of the fixative, or even denaturing of the protein due to how the tissue was processed into sections. Although there are many types of antigen retrieval methods, they all are intended to restore the natural conformation of the target to expose the epitopes. Many IHC protocols provided by the manufacturers of antibodies do not include antigen retrieval and, therefore, in your preliminary test of the antibody you may get no staining at all or very weak staining. Test one or two different antigen retrieval methods in your next experiment while leaving the rest of the protocol the same and your results should be improved.

Antigen retrieval methods can be heat-induced epitope retrieval (HIER), proteolytic-induced epitope retrieval (PIER) or room temperature epitope retrieval (RTER). HIER involves heating your tissue samples to between 95–100 °C in a buffer of a certain pH. The heating can be done in a water bath, microwave, pressure cooker or even an autoclave. These buffers include citrate buffer (pH 6), ethylenediaminetetraacetic acid (EDTA) buffer (pH 8) and Tris buffer (pH 10). PIER uses enzymatic digestion (proteinase K, trypsin, pepsin and others) to break the cross-links, although these methods can cause tissue damage if not optimised for time and temperature. RTER uses acid treatment (2 N hydrochloric acid, pH 0.6–0.9) of the samples at room temperature. As stated earlier, all of these methods aim to break the cross-links made by the fixation process and restore epitope availability to your primary antibody. Good protocols for all of these methods (and a few others) can be found online on antibody manufacturers' websites and at www.ihcworld.com.

9.5.6 Reducing Background Staining

Background staining is one of the bug-bears of IHC methods; it is random, non-specific staining of your section and reducing this is important in producing well-differentiated sections. Let us begin by defining an experiment. We are looking for protein P in sections of human tissue. We have a primary antibody to P. This antibody was raised in a mouse. When we expose the tissue to the antibody, we find a lot of background staining occurs: we would like to reduce this and leave only P stained. We may be identifying our primary antibody by means of a secondary antibody raised in a rabbit. What we need is serum containing antibodies to non-P components of human tissue. This is described as blocking serum. Usually, normal serum or protein solutions are used for this purpose. In practice, it is usual to use a blocking serum from the species used to raise the secondary antibody. The blocking serum will contain antibodies to all sorts of protein and it will bind to all sorts of antigenic sites and prevent our primary antibody (raised in a mouse) binding to anything other than P. Background staining by the primary antibody will have been prevented. The timing and concentration of the blocking serum needs to be titrated: too little blocking and the background will remain high; too much blocking and the epitopes recognised by the primary antibody will be blocked as well. If we had used a blocking serum from a mouse, we would have created a problem. As the animals used to produce antibodies are exposed to thousands of antigens in their lives, the serum will contain the antibodies to these, as well as the antibody to our target protein. The mouse antibodies in the blocking serum could be recognised by the other mouse antibodies in the serum of the primary antibody and background staining would be increased.

Protein solutions, such as bovine serum albumin (0.1–5%), work for this purpose as well by binding to all proteins and causing the primary antibody to out-compete the solution due to its specificity for an epitope. There are also commercially available blocking buffers, which work in a similar fashion, although the exact composition is only known by the company.

If you are using an enzymatic technique for recognising the primary antibody bound to P then controlling the development of the coloured marker will be important in limiting background staining. The objective is to achieve strong colouration of the target without staining of the background. Figure 9.2 shows a well-differentiated section. In all IHC techniques, washing of the sections is critically important: removing residual antibody from the sections is a key step. If you are using a peroxidase-based staining protocol, it is important to determine whether there are peroxidases naturally occurring in your sample. Erythrocytes, liver and kidney all have high levels of these endogenous peroxidases, which result in the decomposition of hydrogen peroxide. Removing this activity is simply achieved by treating your sample with a 3% solution of hydrogen peroxide for 5–10 minutes: all of the endogenous peroxidase activity should be exhausted. A similar process needs to be considered if you are using an alkaline phosphatase method, especially in the kidney, intestine or lymphoid tissues. Treatment of your sample with 5 mM levamisole should exhaust the endogenous phosphatase activity in your sample. If you are unsure whether either of these apply to your samples, adding these treatments in as a matter of course will not affect the outcome of your staining if there is no enzymatic activity in your samples.

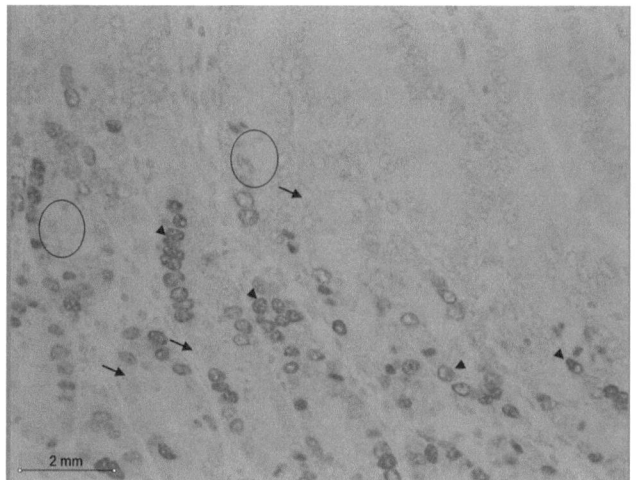

Figure 9.2 **Positive control slide with minimal background staining.** This slide from the duodenum of the rat is showing proliferating cells using bromodeoxyuridine (BRDU). This method of detection has a characteristic nuclear staining pattern, which can be seen in the cells at the base of the crypts (brown nuclei, arrowheads at some examples), while non-proliferating cells have been counter-stained with haematoxylin (blue nuclei, arrows). Background in this sample is at a minimum level and is detectable in the crypts themselves (circled) and has no set pattern, is not in the nuclei of the cells and does not obscure the signal from the positive cells.
This example was graciously supplied by K. Belsham, Sequani Ltd.

Some antibody optimisation troubleshooting tips are shown in Table 9.1.

9.5.7 Amplification kits

As mentioned earlier, it is possible to buy kits that can amplify the signal resulting from your antibody detection. These kits are useful when you are unsure of the level of expression of your target protein, when you know that the level is likely to be quite low or, due to various constraints, you have to use a minimum of primary antibody on your samples. An example of signal amplification is shown in Figure 9.1D with the use of an avidin–biotin reaction. The amplification of the signal is due to the interaction of multiple avidin–biotin complexes to the biotin attached to the secondary antibody, which can then produce a larger reaction to the horseradish peroxidase. Further modifications to this have been developed for even greater sensitivity.

Table 9.1 Troubleshooting during antibody optimisation.

Problem	Check 1st	Check 2nd
No Staining/ faint staining	Check expiry dates on all solutions	Try different antigen retrieval methods: see section 9.5.5 for options
	Ensure you have done dilutions correctly	Decrease the concentration or timing of blocking
	Check that your samples have been fixed and processed correctly for the antibody (paraffin vs. frozen etc.)	Do a dilution series with your primary antibody: increase antibody concentration
	Ensure you have the correct positive control tissue	If have using paraffin sections and have access to frozen sections, try frozen sections
	Ensure that your secondary should recognise the species of the primary	Check adequacy of fixation: see Chapter 5
		Increase time for colour development
High background	Check the species of your primary antibody	Decrease concentration of secondary antibody
	Ensure you have carried out an endogenous block, if necessary	Increase time of washing step
	Decrease the concentration of primary antibody	Increase concentration or timing of blocking
Over-staining	Ensure sections did not dry out at any point during the protocol	Decrease time of colour development
	Decrease concentration of primary antibody	Decrease incubation time and/or temperature

These other methods, which do not involve avidin–biotin and thus avoid background caused by endogenous biotin, are generally patented by individual suppliers. However, as an example, Dako[3] produces a kit called EnVision™, which, instead of using a standard secondary antibody, uses a dextran backbone with up to ten secondary and up to seventy enzyme molecules attached. This results in a great deal of amplification of the primary signal. (See the Dako website and handbook for further information.)

9.5.8 Using IHC as an Experimental Tool

Given all of the considerations above, the reader may well think that IHC is a complicated technique and one which he or she might, perhaps, avoid. This would be an error: IHC methods are not difficult; they are fiddly but, once mastered, excellent results can be produced on a routine basis. An example will help.

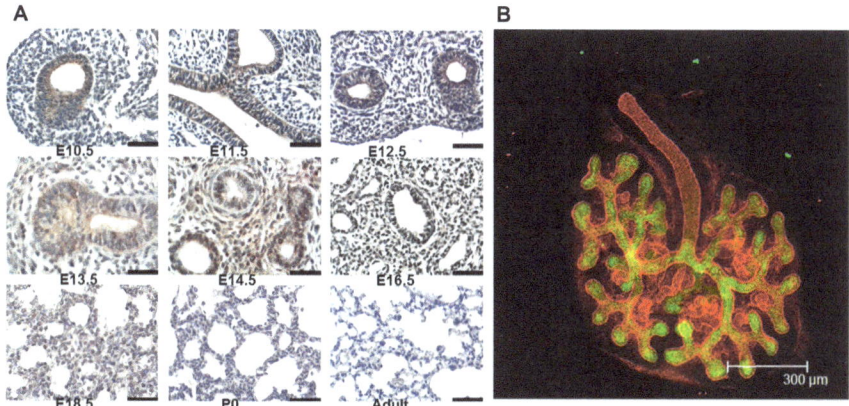

Figure 9.3 **Experiments using immunohistochemistry. A.** Localisation of the calcium-sensing receptor determined by IHC over the course of lung development shows that it is localised initially to the epithelium before also being expressed in the mesenchyme, with expression being lost at birth. **B.** Confocal imaging stack through the depth of a cultured mouse embryonic kidney stained with the basement membrane protein laminin (red) and calbindin (green) for counting the branching morphogenesis, which occurs in culture.
Image taken by B. Finney at Cardiff University, Molecular Sensing Group Laboratory.

We can use Finney *et al.* (2008)[1] as an example from published literature of the usefulness of IHC technique. A polymerase chain reaction (PCR) showed the presence of the extracellular calcium-sensing receptor (CaR) in developing mouse lungs from embryonic day 10.5 to 18.5, but this did not give any indication of the cell type expressing this receptor, or the function of the CaR in this context. IHC was performed using a commercially available antibody that recognised an amino acid sequence specific to the CaR on paraffin wax sections of developing lung (Figure 3A). This showed that the receptor is initially expressed in the epithelium, but as development progresses expression begins to appear in the mesenchyme as well. This changing expression pattern provided insights into the potential role of the CaR in the developing lung and guided further work in this area. The basic methods used are set out below.

9.5.9   Basic Immunohistochemistry Protocol: Using Horseradish Peroxidise and DAB Detection

Solution

Phosphate-buffered saline (PBS), pH 7.4.

Method

1. De-wax and rehydrate slides:

Histochoice®/xylene	2 minutes
Histochoice®/xylene	2 minutes
100% ethanol	3 minutes
90% ethanol	30 seconds
75% ethanol	30 seconds
50% ethanol	30 seconds
30% ethanol	30 seconds
Distilled water	3–5 minutes

2. Antigen retrieval:

   Place slides in microwave-safe container and cover with citrate buffer, pH 6.

   Microwave at 900 W for 3–4 minutes until solution is boiling.

   Microwave at 300 W for 10 minutes (ensure slides remain covered by solution the entire time).

   Allow to cool to room temperature.

3. Blocking:

   Make up blocking solution: 3–5% serum + 0.1% Triton-X 100 in PBS.

   Mix well.

   Cover each tissue section with blocking solution.

   Incubate for 1 h at room temperature.

4. Endogenous peroxidase block (this can be done here or after step 5):

   Tap off blocking solution on to tissue.

   Add drops of endogenous peroxidise block (hydrogen peroxide solution or commercially available mix).

   Leave for 5–10 minutes at room temperature.

5. Apply primary antibody:

   Make up antibody dilution fluid: 3% serum (same as blocking step) + 0.1% Triton-X 100 in PBS.

   Dilute antibody 1:200; for negative controls, dilute irrelevant IgG at the same concentration.

   Cover tissue sections with primary antibody.

   Leave at room temperature for 1 h, or overnight at 4 °C.

6. Wash:

   Tap off primary antibody solution onto tissue (in some cases, you can remove with a pipette and save for the next experiment).

   Wash 3×5 minutes in PBS.

7. Secondary antibody:

Add horseradish peroxidise-labelled secondary antibody at concentration recommended by manufacturer in 1% bovine serum albumin + 0.1% Triton-X 100 in PBS.

Incubate at room temperature for 30 minutes to 1 h.

8. Wash:

Tap off antibody solution onto tissue.

Wash 3×5 minutes in PBS.

9. Visualisation:

Make up substrate chromagen and DAB according to manufacturer's instructions; tap off wash solution.

Put positive control slide on microscope then apply DAB solution, time the colour development with a stopwatch.

After sufficient colour development, rinse off DAB with PBS using a squirt bottle; place the slide into PBS.

Develop the negative control slide for the same amount of time as the positive control.

10. Counterstain and dehydrate:

Dip slide in Harris' haematoxylin (5–20 dips depending on the colour wanted).

Rinse under flowing tap water.

Dip in acid alcohol if the colour is too dark.

Blue slides in Scott's tap water if necessary.

Dehydrate by going through a reverse gradient of ethanols to Histochoice/xylene:

75% ethanol	3 minutes
90% ethanol	3 minutes
100% ethanol	3 minutes
Histochoice/xylene	5 minutes, or as long as possible in Histochoice/xylene

11. Mounting/coverslipping:

Dry off slides with tissue, without touching tissue sample.

Add two drops of the mounting medium onto tissue sample.

Slowly lower the coverslip; remove any excess mounting medium and bubbles.

Leave to set.

9.5.10 Other Useful IHC Markers

It can be a daunting task to wade through the literature to find the right markers to add depth and quality to your research study. As an aid to helping you, we have provided the following table (Table 9.2) as

Table 9.2 Commonly used antibodies.

Purpose	Antibody	Target detected
Proliferation	PCNA	Proliferating cell nuclear antigen
	Phospho-histone H3	Phosphorylated histone H3
	Ki-67	KI-67
	BrdU*	Bromodeoxyuridine
Apoptosis	Cleaved caspase-3	Active form of caspase-3
	Bcl-2	B-cell lymphoma 2, regulator protein
	TUNEL*	Terminal deoxynucleotidyltransferase dUTP nick-end labelling, labels fragmenting DNA
Blood vessels	VE-cadherin	Vascular endothelial cadherin/CD144, blood vessel adhesion molecule
	Endomucin	An endothelial sialomucin found on capillaries and venules. Not useful for arterioles
	PECAM	Platelet endothelial cell adhesion molecule/CD31, endothelial cell intracellular junctions
Hormone production	FSH	Follicle stimulating hormone – pituitary gland
	LH	Luteinising hormone – pituitary gland
	Insulin	Pancreas – islets of Langerhans
	Glucagon	Pancreas
Immunophenotyping (CD, cluster of differentiation markers)[a]	CD3	T-lymphocytes
	CD11c	Leukocytes
	CD45	Leukocytes
	CD34	Haematopoetic stem cells
	CD63	Activated basophils
	CD117	c-kit/stem cell growth factor receptor
*BRDU	Animals must be administered with BRDU prior to euthanasia, which binds in the nucleus of proliferating cells and is then detected by an antibody	Can be administered to pregnant females to detect proliferation in the foetus
*TUNEL	Not an antibody as we have described, usually sold in kit form with all necessary components and the protocol is similar to an IHC process	

[a]There are over 300 of these; the combinations that are expressed on a cell can help determine its identity. Used in immunology and cancer diagnosis.

a short guide to commonly used antibodies with their purpose and target. This list is by no means exhaustive, new antibodies are being made and reported all the time. Take the time to read the literature associated with your target and this will give you a good idea as to the standard reagents. If you are unlucky (or truly pursuing something unique!) you may have to start from scratch, but hopefully this chapter has helped you with your background IHC knowledge and you can hit the ground running once you have made/sourced your new antibody.

As a note of caution, look for other names of your protein of interest. You may read a paper where it says that X is a specific marker for Z, but if X is also known as Q in another discipline, it may also be reported as a specific marker for something completely different! For example, podoplanin was reported to be specific for kidney podocytes but, known as T1α, it was reported as a marker for lung alveolar type 1 cells, as GP38 for hepatic progenitor cells and it can also be found up-regulated in some tumours. So, make sure you do a thorough background search: it may save you a significant amount of time in proving the specificity of your antibody choice.

REFERENCES

1. B. Finney, P. M. delMoral, W. J. Wilkinson, S. Cayzac, M. Cole, D. Warburton, P. J. Kemp and D. Riccardi, *J. Physiol.*, 2008, **15**(586), 6007–6019.
2. J. W. Lichtman and J. A. Conchello, *Nat. Methods*, 2005, **2**, 909–919.
3. Dako, www.Dako.com, Accessed and Downloaded 07-10-2013, for information regarding the EnVision™ system, as well as their excellent IHC handbook.

Appendix

Formulae or recipes not set out in full in the text

Formol saline (formal-saline)
Formalin (the 40% solution of formaldehyde in water): 100 mL
Sodium chloride: 8.5 g
Tap water: 900 mL
The mixture tends to become acid and a layer of marble chips (calcium carbonate) used to be placed at the bottom of the bottle. Buffered formalin provides better control of pH.

Buffered formalin (most widely used fixative solution)
Formalin (the 40% solution of formaldehyde in water): 100 mL
Tap water: 900 mL
Sodium dihydrogen orthophosphate, $Na_2H_2PO_4 \cdot H_2O$: 4 g
Disodium hydrogen orthophosphate, anhydrous, Na_2HPO_4: 6.5 g
This mixture has a pH of 7.0.

Bouin's fixative
Saturated aqueous solution of picric acid: 75 mL
Formalin (the 40% solution of formaldehyde in water): 25 mL
Acetic acid: 5 mL (this is, of course, glacial acetic acid)
Keeps well. The saturated picric acid can be decanted from a jar of picric acid in which a layer of picric acid is kept under water. Remember to replace the water.

Histological Techniques: An Introduction for Beginners in Toxicology
By Robert Maynard, Noel Downes and Brenda Finney
© Maynard, Downes and Finney 2020
Published by the Royal Society of Chemistry, www.rsc.org

Zenker's fixative
   Mercuric chloride: 5 g
   Potassium dichromate: 2.5 g
   Sodium sulphate: 1 g
   Distilled water: 100 mL
   Acetic acid: 5 mL, to be added just before you use the fixative
The mixture, without the acetic acid, keeps well: as soon as the acetic acid is added its life is limited.

Zenker formol (Zenker-formal, Helly's fluid)
   This is exactly the same as Zenker's fixative, except that the acetic acid is replaced with 5 mL formalin (the 40% solution of formaldehyde in water) immediately before use.

Susa fixative
   Mercuric chloride: 45 g
   Sodium chloride: 5 g
   Trichloracetic acid: 20 g
   Acetic acid: 40 mL
   Formalin (the 40% solution of formaldehyde in water): 200 mL
   Distilled water: 800 mL

Mercuric chloride–formalin
   Saturated solution of mercuric chloride in distilled water: 90 mL
   Formalin (the 40% solution of formaldehyde in water): 10 mL

Carnoy's fluid
   Absolute alcohol: 60 mL
   Chloroform: 30 mL
   Acetic acid: 10 mL

Sanfelice's fluid
   Solution A: formalin (the 40% solution of formaldehyde in water, 128 mL), acetic acid (16 mL)
   Solution B: 1% chromic acid (100 mL)
   Mix 9 mL solution A with 16 mL solution B immediately before you use the fluid.

Flemming's fluid
   1% chromic acid: 15 mL
   2% osmium tetroxide: 4 mL
   Acetic acid: 1 mL
   Keeps badly: make it up as you need it.

Scott's tap water substitute
   Sodium bicarbonate: 3.5 g
   Magnesium sulphate: 20 g
   Tap water: 1000 mL
   Add a crystal of thymol to prevent mold growth.

Annotated Bibliography

I have listed only those books with which I am personally familiar. No doubt there are others that might have been mentioned. I have not hesitated to list books from the golden age of histological technique, though some are now rather old and thus only available second-hand or from better libraries. Most of the works listed provide references to the original literature. Although electron microscopy has not been discussed in this book, I have included some books on electron microscopy in this bibliography. These may be helpful to those wishing to relate light and electron microscopic appearances of tissues.

HISTOLOGICAL TECHNIQUES

Baker J R, *Principles of Biological Microtechnique; a Study of Fixation and Dyeing*, Methuen and Co. Ltd, London, 1958. This and Baker's smaller book, Cytological Technique, are essential reading for anyone wishing to understand how histological techniques work. The appendix on spelling is a real gem.

Baker J R, *Cytological Technique*, Chapman and Hall, London, 5th edn, 1966.

Bancroft J D and Gamble M, *Theory and Practice of Histological Techniques*, Churchill Livingstone, Edinburgh, 6th edn, 2007. Excellent, modern and comprehensive. Widely used in histology laboratories.

Histological Techniques: An Introduction for Beginners in Toxicology
By Robert Maynard, Noel Downes and Brenda Finney
© Maynard, Downes and Finney 2020
Published by the Royal Society of Chemistry, www.rsc.org

Bolles Lee A, *The Microtomist's Vade-Mecum*, J & A Churchill Ltd, London, 4th edn, 1896 (1st edn: 1885). Perhaps THE classic work on histological technique. This edition was written by Bolles Lee himself: every method described in the book was tried by the author: if Bolles Lee said that a technique would work then it will work! Superbly funny in places.

Bracegirdle B, *A History of Microtechnique*, Heinemann, London, 1978. A valuable account of the history of the subject based on the author's doctoral thesis. Especially good on the development of microtomes.

Buchwalow I B and Böcker W, *Immunohistochemistry: Basics and Methods*, Springer-Verlag, Berlin Heidelberg, 2010. A good overview of the subject with protocols following-up each introduction of a method.

Carleton H M and Drury R A B, *Histological Technique*, Oxford University Press, London, 3rd edn, 1957. The last edition prepared by the original author (HMC). Contains a large number of tips and dodges. Charming style. The section dealing with techniques on a tissue-by-tissue basis does not appear in later editions.

Clark G, *Staining Procedures used by the Biological Staining Commission*, Williams & Wilkins Company, Baltimore, 3rd edn, 1973. A compendium of well-tested methods: very reliable.

Clayden E C, *Practical Section Cutting and Staining*, J & A Churchill Ltd, London, 3rd edn, 1955. A very useful little handbook full of practical tips.

Cowdry E V, *Microscopic Technique in Biology and Medicine*, Williams and Wilkins Company, Baltimore, 1943. Arranged as a dictionary or encyclopaedia and packs a great deal of information in a condensed format. Valuable as a reference for out of the way methods.

Davenport H A, *Histological and Histochemical Technics*, W B Saunders Company, 1960. An outstanding practical manual with the best accounts of section cutting techniques that I have come across. Davenport's special expertise in neuro-pathology makes the sections of celloidin methods and silver impregnation especially valuable.

Drury R A B and Wallington E A, *Carleton's Histological Technique*, Oxford University Press, Oxford, 5th edn, 1980. Updated edition of an old classic. Useful for newer methods. Very reliable.

Gabe M, *Histological Techniques*, trans. R E Blackwith and A Kovoor, Masson and Springer-Verlag, Paris and New York, 1976. Unique! Over one thousand pages of advice—much of it expressing the author's personal and often iconoclastic opinions. Always worth consulting. No bibliography.

Gatenby J B, *The Microtomist's Vade-Mecum (Bolles Lee)*, J & A Churchill Ltd, London, 8th edn, 1921. Later edition of a classic.

Gatenby J B, *The Microtomist's Vade-Mecum (Bolles Lee)*, J & A Churchill Ltd, London, 10th edn, 1937. Penultimate edition of a classic. Larger than previous editions. This and the 11th edition are still referred to as definitive sources of advice. One annoying feature is the inclusion of references to earlier editions ("for...see 7th edition..."): I have three; who has all eleven?

Gomori G, *Microscopic Histochemistry*, University of Chicago Press, Chicago, 1952. An outstanding early account.

Gray P, *The Microtomist's Formulary and Guide*, Robert E Kreiger Publishing Company, New York, 1954. Reprinted 1975. Like Gabe's book, this is a unique work. Nobody else has taken on the task of listing the many thousands of histological techniques that have been developed. A catalogue and an encyclopaedia. The author's account of how to prepare a mid-line sagittal section of an entire mouse, including every vertebra of the tail, should be read by all budding histology technicians.

Gurr E, *A Practical Manual of Medical and Biological Staining Methods*, Leonard Hill [Books] Ltd, London, 1956. The first of Edward Gurr's four books. Very practical and reliable. The book is an expanded collection of the pamphlets on staining produced by Edward Gurr just after the Second World War.

Gurr E, *Methods of Analytical Histology and Histo-chemistry*, Leonard Hill [Books] Ltd, London, 1958. Another valuable book by Edward Gurr, in which his practical experience shines through.

Gurr E, *Encyclopaedia of Microscopic Stains*, Leonard Hill [Books] Ltd, London, 1960. A very valuable work. The information given on many of the dyes included is difficult to find elsewhere. Particularly valuable for notes on the use of dyes in countries other than the UK.

Gurr E, *Synthetic Dyes*, Academic Press, London, 1971. Provides the absorption curves for each dye listed. Helpful when used in parallel with Conn's *Biological Stains*. Gurr used his own numbering system for dyes. He pointed out that the colour index numbers changed from time to time but that his didn't.

Gurr G T, *Biological Staining Methods*, Searle Scientific Services, High Wycombe, England, 7th edn, 1969. A useful little manual by Edward Gurr's brother.

Hall J W and Herxheimer G, *Methods of Morbid Histology and Chemical Pathology*, William Green and Sons, Edinburgh and London, 1905. Classic work.

Kiernan J A, *Histological and Histochemical Methods, Theory and Practice*, Scion Publishing, 4th edn, 2008. In depth chemical background of histochemical methods with clear recipes and protocols.

Lillie R D, *H J Conn's Biological Stains*, Williams & Wilkins Company, Baltimore, 8th edn, 1969. Essential for reference. Later editions are available. Provides the colour index reference numbers, which are missing in Gurr's books.

Lillie R D and Fullmer H M, *Histopathologic Technic and Practical Histochemistry*, McGraw-Hill Book Company, New York, 4th edn, 1976. The best of all books on techniques and their theoretical background. Praised for its description of histochemical methods by no lesser an authority than A G E Pearse.

Mallory F B, *Pathological Technique*, W B Saunders Company, Philadelphia and London, 1938. Useful older account by an expert.

Mann G, *Physiological Histology*, Clarendon Press, Oxford, 1902. A very important early work: ahead of its time. Mann sought to explain the theory of histological techniques in physico-chemical terms and succeeded brilliantly given the period in which he wrote. Still well worth reading in parts but not easy going. Rather rare today.

Mayer E, *Introduction to Dynamic Morphology*, Academic Press, New York and London, 1963. A most important work and the only one of its kind. Explains morphological research methods in a way not found in any other book. Anyone who uses morphological techniques should refer to this book. Brilliant on artefacts. Seems to be rather rare.

McClung Jones R, *McClung's Handbook of Microscopical Technique*, Cassell and Company Ltd, London, 3rd edn, 1950. A classic work.

McClung Jones R, *Basic Microscopic Technics*, University of Chicago Press, Chicago and London, 1966. Invaluable for detailed information on apparatus and equipment.

Pearse A G E, *Histochemistry, Theoretical and Applied*. Churchill Livingstone, Edinburgh, London and New York, 3rd edn, 1968. A classic work.

Pantin C F A, *Notes on Microscopical Technique for Zoologists*, Cambridge University Press, Cambridge, 1964. A very useful little manual of techniques.

Polak J M and Van Noorden S, *Introduction to Immunocytochemistry*, BIOS Scientific Publishers Ltd, Oxford, 3rd edn, 2003. An excellent and up to date introduction. Includes an attractive set of coloured plates.

Squire P W, *Methods and Formulae Used in the Preparation of Animal and Vegetable Tissues for Microscopical Examination*, J & A Churchill, London, 1892. Included here as one of the oldest formularies in the field. Referred to by Bolles Lee.

MICROSCOPY

Barer R, *Lecture Notes on the Use of the Microscope*, Blackwell Scientific Publications, Oxford and Edinburgh, 3rd edn, 1968. The best of all short accounts by a master microscopist. Barer held degrees in physics, physiology and medicine.

Beck C, *The Microscope*, R & J Beck Ltd, London, 1938. Excellent account of the classical theory.

Bradbury S, *The Evolution of the Microscope*, Pergamon Press, Oxford, 1967. A historical survey of real distinction by a leading British authority.

Bradbury S, 'The Optical Microscope in Biology', in *Institute of Biology's Studies on Biology*, no. 59, Edward Arnold, London, 1976. Most helpful short account.

Bradbury S, 'An Introduction to the Optical Microscope', in *Royal Microscopical Society: Microscopy Handbook*, 01, Oxford University Press, Oxford, 1984. Another short but very valuable account by the same author as above.

Bradbury S, Evennett P J, Haselmann H and Piller H, 'Dictionary of Light Microscopy', in *Royal Microscopical Society: Microscopy Handbooks*, 15, Oxford University Press, Oxford, 1989. Invaluable.

Burrells W, *Microscopic Technique*, Fountain Press, London, revised edition, 1977. Wide-ranging account of methods: I have not found it as helpful as Barer's much shorter book. Useful for out of the way information.

Cox A, *Photographic Optics*, Focal Press, London and New York, 13th edn, 1966. More recent editions are available. Though aimed at the photographer, this book contains much of interest to the microscopist. The discussion of defects of lenses is masterly.

Delly J G, *Photography Through the Microscope*, Eastman Kodak Company, Rochester, 7th edn, 1980. This is the best modern account of how to set up a microscope that I have come across. The theoretical background is explained in detail. The account owes a good deal to Shillaber.

Gage S H, *The Microscope*, Constable and Company Ltd, 17th edn, London, 1943. A classic account.

Hartley W G, *Hartley's Microscopy*, Senecio Publishing Company Ltd, Charlbury, 2nd edn, 1979. A useful introduction. My copy has pages missing (133–148) due, presumably, to an error in binding.

Shillaber C P, *Photomicrography in Theory and Practice*, John Wiley and Sons Inc., New York and Chapman and Hall Ltd, London, 1947.

The best big book on the subject. Should be read by all that use the light microscope. Do not be put off by its age.

Stephens D, *Cell Imaging*, Scion Publishing, 2006. Further in-depth information on fluorescence microscopy techniques and advanced technology.

HISTOLOGY

Histology used to form a substantial part of the medical curriculum and was well served by large, regularly revised textbooks. Times have changed and the large works listed below have largely disappeared. Copies no doubt linger on in the basements of medical school and university libraries. Out of print texts can often be bought second-hand: the various Internet book-search engines allow copies to be tracked down and bought with minimal effort and at modest expense.

Modern Classics of Human Histology

Fawcett D W, *Textbook of Histology*, Chapman and Hall, New York and London, 12th edn, 1994. This seems to be the final edition of perhaps the most famous of all textbooks of histology. The first edition was published in 1930 and was known as "Maximow and Bloom", then, for many years, as "Bloom and Fawcett". Detailed, authoritative, indispensable. Illustrations have changed from edition to edition: earlier editions have more of Maximow's outstanding drawings and more photomicrographs taken using light microscopy; in later editions more electron micrographs appear.

Weiss L, *Histology, Cell and Tissue Biology*, Macmillan Press, London, 5th edn, 1983. Another large book with a long history: F T Lewis' translation of Stöhr's *Lerbuch der Histologie* reached its 5th edition under Bremer and Weatherford in 1944; the book was later taken over and much enlarged by Greep and then by Weiss. The 1983 edition, the 5th, is particularly strong on electron microscopy. There is a 6th edition.

Ham A W, *Histology*, J B Lippincott Company, Philadelphia and Toronto, 7th edn, 1974. Last large edition of "Ham": later editions are less detailed. More "user friendly", if rather more discursive, than Bloom and Fawcett; excellent on Ham's own area of research: bone and cartilage.

Rhodin J A G, *Histology, A Text and Atlas*, Oxford University Press, London and Toronto, 1974. The first text to focus on electron microscopy. Illustrations produced as an atlas: see the section on Atlases of Histology. A fine and very widely quoted work.

Clark W E Le Gros, *The Tissues of the Body*, Clarendon Press, Oxford, 6th edn, 1975. One of the truly great books on histology of the 20th century. Recommended to all histologists.

Older Classics

Cowdry E V, *General Cytology*, University of Chicago Press, Chicago, Illinois, 1925. A fine work summarising much of what was known of cytology at the time of publication.

Cowdry E V, *Special Cytology*, Paul E Hoeber Inc., New York, 2nd edn, 3 vols, 1932. Tissue-by-tissue account, distinguished contributors, indispensable for detailed explanations of light microscopic appearances and for early references. Difficult to obtain.

Cowdry E V, *A Textbook of Histology: Functional Significance of Cells and Intercellular Substances*, Lea and Febiger, Philadelphia, 2nd edn, 1938. Most interesting attempt to introduce a discussion of functional aspects of histology. Later editions lost some of Cowdry's verve.

Wilson E B, *The Cell in Development and Heredity*, The Macmillan Company, New York, 3rd edn, 1928. The last and best edition of this rightly famous classic. All histologists should at least leaf through a copy. Invaluable for references to the earlier literature.

Brachet J and Mirsky A E, *The Cell, Biochemistry, Physiology, Morphology*, Academic Press, New York and London, 1959. Six large volumes: the last appearing in 1964. Outstandingly good accounts of all aspects of cytology as understood in the early 1960s.

Willmer E N, *Cells and Tissues in Culture*, Academic Press, London and New York, 3 vols, 1965. Covers all aspects of cell culture as practised in the 1960s. Superb accounts of individual cell types: see, for example, Jacoby's account of the macrophage. Valuable for references to the older literature.

Mottram V H, *A Manual of Histology*, Methuen & Co. Ltd, London, 1923. Now rather rare. This work contains 224 pen and ink drawings made and signed by the author himself. Of course the text, now 90 years old, is dated but the advice "to the student" is as fresh as if it had been written today. The student or trainee in toxicology might be encouraged by the author's concluding remarks "to the student": *"Histology is not difficult, it merely has endless detail and is tedious. Anyone can master histology who has patience and perseverance. It will be of the greatest use to the student if he will master it early in his career, and remember not to forget it."* Professor V H Mottram was a distinguished physiologist and the father of Dr R F Mottram: teacher, colleague and friend of the present author.

Atlases of Histology

Many atlases have appeared over the years. Some are illustrated with drawings and paintings, other with photomicrographs.

Craigmyle M B L, *A Colour Atlas of Histology*, Wolfe Medical Publications Ltd, London, 2nd edn, 1986. A collection of colour photographs. Excellent, with short descriptions of each section. Human tissue only. Sections prepared by Mr Les Jones (Chief Technician, Department of Anatomy, University College Cardiff), who taught the present author about histological techniques.

Di Fiore M S H, *An Atlas of Human Histology*, Henry Kimpton, London, 3rd edn, 1967. Coloured drawings of a high standard.

Burkitt H G, Young B and Heath J W, *Wheater's Functional Histology, A Text and Colour Atlas*, Churchill Livingstone, Edinburgh, 3rd edn, 1993. Easily the best and most popular atlas now available. Used as a textbook in many medical schools. Later editions are available.

Hammersen F, *Histology, A Colour Atlas of Cytology, Histology and Microscopic Anatomy*, Urban & Schwarzenberg, Baltimore-Munich, 2nd edn, 1980. A much updated edition of Sobotta's atlas of 1911. Very high quality photographs, including electron microscopy.

Krstic R V, *Illustrated Encyclopedia of Human Histology*, Springer-Verlag, Berlin, 1984. A unique work, including a huge amount of detailed information and illustrated with the author's drawings showing the three dimensional structure of tissues. All histologists should have a copy for reference.

Atlases and Textbooks Focusing on Electron Microscopy

Rhodin's book on histology might have been included here; the illustrations from his book (see above) were collected in an atlas, which is listed here.

Carr K E and Toner P G, *Cell Structure (An Introduction to Biomedical Electron Microscopy)*, Churchill Livingstone, Edinburgh, 3rd edn, 1982. A very valuable account with an atlas of fine electron micrographs.

Threadgold L T, *The Ultrastructure of the Animal Cell,* Pergamon Press, Oxford, 2nd edn, 1976. A detailed account, which is still very useful.

Fawcett D W, *The Cell. An Atlas of Fine Structure*, W B Saunders Company, Philadelphia and London, 1966. The first great atlas of fine structure. A second edition was produced.

Rhodin J A G, *An Atlas of Histology*, Oxford University Press, New York, 1975. The illustrations are reproduced in a larger format than in Rhodin's *Histology* and on better quality paper. Particularly valuable for the low power electron microscopy (EM) pictures and the light microscopy (LM) pictures of thin sections. A very important collection of photographs. Perhaps becoming rather rare.

Comparative Histology

Textbooks of comparative histology are few. Textbooks for veterinary students deal with domestic animals but no really great textbook of comparative histology has yet been written, at least not in English. This point was made by J Z Young in the 1950s in the preface to his *The Life of Vertebrates* but nobody seems to have responded to his suggestion that such a work was needed.

Dellman H D and Eurell J A, *Textbook of Veterinary Histology*, Lippincott, Williams and Wilkins, Philadelphia, 5th edn, 1998. One of several excellent books aimed at veterinary students.

Dahlgren U and Kepner W A, *Principles of Animal Histology*, The Macmillan Company, New York, 1908. (My copy is dated 1925, presumably a reprint; no second edition seems to have been published.) A truly comparative account arranged on a tissue-by-tissue basis and including invertebrates. Useful, though over one hundred years old.

Welsch U and Storch V, *Comparative Animal Cytology and Histology*, Sidgwick & Jackson, London, 1976. In some ways, this is almost a modern version of Dahlgren's *Principles of Animal Histology*. Excellent text and figures.

Patt D I and Patt G R, *Comparative Vertebrate Histology*, Harper Row, New York, Evanston and London, 1969. A useful, modern and short account.

Andrew W, *Textbook of Comparative Histology*, Oxford University Press, New York, 1959. A thoughtful account with study questions. Readable style.

Kendall J I, *Microscopic Anatomy of Vertebrates*, Lea and Febiger, Philadelphia, 3rd edn, 1947. Old and dry but useful; draws on Maximow and Bloom for human histology.

Hodges R D, *The Histology of the Fowl*, Academic Press, London, 1974. A detailed account of the histology of a non-mammal! Histologists and immunologists often speak of the Bursa of Fabricius: Hodges provides a suitably detailed account.

PATHOLOGY

There is no shortage of books on human and veterinary pathology. Books aimed at medical and veterinary students are updated regularly. Here, a selection of new and older textbooks and atlases is provided. Readers will, no doubt, have other favourites. In each case the latest edition should, of course, be consulted. Works dealing with the pathology of specific organs have not been included.

Kumar V, Abbas A K and Aster J, *Robbins and Cotran Pathologic Basis of Disease*, W B Saunders Company, Philadelphia, 8th edn, 2010. Perhaps the best of all current textbooks of human pathology. Includes a great deal of molecular biology. A shorter version is also available.

Sternberg S S, *Histology for Pathologists*, Raven Press, New York, 1992. Best of all books on histology for the pathologist: excellent text and superb colour illustrations throughout. A second edition is available.

Danjanov I and Linder J, *Anderson's Pathology*, C V Mosby Company, St Louis, 10th edn, 2 vols, 1996. Detailed account of human pathology.

Benirschke K, Garner F M and Jones T C, *Pathology of Laboratory Animals*, Springer-Verlag, New York, 2 vols, 1978. Indispensable to the diagnostic pathologist working with laboratory animals.

Florey H W, *General Pathology*, Lloyd-Luke (Medical Books) Ltd, London, 4th edn, 1970. A classic work with contributions from particularly distinguished authors. Old but still valuable. Florey's account of the inflammation of mucous surfaces is masterly. Florey brought a physiologist's approach to pathology; he also won the Nobel Prize, with Flemming and Chain, for work on penicillin.

Ghadially F N, *Ultrastructural Pathology of the Cell*, Hodder-Arnold, England, 4th edn, 2 vols, 1997. An atlas and a textbook: uniquely detailed. Invaluable for those interested in pathological changes at the EM level.

Cameron G R, *Pathology of the Cell*, Oliver and Boyd, Edinburgh, 1952. A classic work, more admired than read. Valuable for its historical approach and the thousands of references to the older literature.

Curran R C, *Colour Atlas of Histopathology*, Oxford University Press, London, 4th edn, 2000. An outstanding collection of colour photographs with accompanying notes.

ORGANIC CHEMISTRY

Knowledge of histological techniques requires some familiarity with organic chemistry and with the chemistry of dyes. For most histologists, a limited familiarity serves. The following books have proved consistently useful to the author.

Brown G I, *Introduction to Organic Chemistry*, Longmans, Green and Co. Ltd, London, 1957. A school textbook that provides a short but useful section on dyes and a lot of basic information.

Karrer P, *Organic Chemistry*, H V Simon and N G Bisset (eds), Elsevier Publishing Company Inc., New York and London, 4th edn, 1950. A large and perhaps old fashioned book of nearly one thousand pages. Very helpful for looking up formulae; good section on dyes.

Finar I L, *Organic Chemistry*, Longmans, London, vols 1 and 2, 5th edn, 1967. A large, more modern text that adopts a mechanistic approach not found in Karrer's *Organic Chemistry*. The section on dyes is most helpful and the explanation of why dyes are coloured will delight the chemically sophisticated reader.

Parry E J, *The Chemistry of Essential Oils and Artificial Perfumes, Volume 1: Monographs on Essential Oils*. Scott, Greenwood and Son, London, 1921. A fascinating old book containing a wealth of information on all the essential oils then known. Emphasis on botany and production methods. Very valuable when thinking about clearing agents.

Thorpe J F and Whiteley M A, *Thorpe's Dictionary of Applied Chemistry*, Longmans, Green and Co., London, 11 vols (index vol. as vol. 12), 4th edn, 1937–1954. A vast compendium of information on all aspects of applied chemistry and its history, which is still found in better libraries. Bound in cloth stained with Monastral Fast Blue, BS, referred to by the editors as: "the greatest chemical achievement of 1936"! The history of the discovery of the dye is given in the preliminary pages of volume 1. Worth tracking down in a library before the remaining copies are discarded.

Subject Index

Page numbers in *italics* refer to figures. The suffix T indicates a table.

£40.00

Lightning Source UK Ltd.
Milton Keynes UK
UKHW050621111219
355150UK00004B/106/P

9 781839 161476